The approved documents

What is an approved document?

The Secretary of State has approved a series of documents that give practical guidance about how to meet the requirements of the Building Regulations 2010 for England. These approved documents give guidance on each of the technical parts of the regulations and on regulation 7 (see the back of this document). The approved documents provide guidance for common building situations.

It is the responsibility of those carrying out building works to meet the requirements of the Buildings Regulations 2010. Although it is ultimately for the courts to determine whether those requirements have been met, the approved documents provide practical guidance on potential ways to achieve compliance with the requirements of the regulations in England.

Although approved documents cover common building situations, compliance with the guidance set out in the approved documents does not provide a guarantee of compliance with the requirements of the regulations because the approved documents cannot cater for all circumstances, variations and innovations. Those with responsibility for meeting the requirements of the regulations will need to consider for themselves whether following the guidance in the approved documents is likely to meet those requirements in the particular circumstances of their case.

Note that there may be other ways to comply with the requirements than the methods described in an approved document. If you prefer to meet a relevant requirement in some other way than that described in an approved document, you should seek to agree this with the relevant building control body at an early stage.

Where the guidance in the approved document has been followed, a court or inspector will tend to find that there is no breach of the regulations. However, where the guidance in the approved document has not been followed, this may be relied upon as tending to establish breach of the regulations and, in such circumstances, the person carrying out building works should demonstrate that the requirements of the regulations have been complied with by some other acceptable means or method.

In addition to guidance, some approved documents include provisions that must be followed exactly, as required by regulations or where methods of test or calculation have been prescribed by the Secretary of State.

Each approved document relates only to the particular requirements of the Building Regulations 2010 that the document addresses. However, building work must also comply with all other applicable requirements of the Building Regulations 2010 and all other applicable legislation.

How to use this approved document

This document uses the following conventions.

a. Text against a green background is an extract from the Building Regulations 2010 or the Building (Approved Inspectors etc.) Regulations 2010 (both as amended). These extracts set out the legal requirements of the regulations.

b. Key terms, printed in green, are defined in Appendix A.

c. References are made to appropriate standards or other documents, which can provide further useful guidance. When this approved document refers to a named standard or other reference document, the standard or reference document has been clearly identified in this document. Standards are highlighted in **bold** throughout. The full name and version of the document referred to is listed in Appendix F (standards) or Appendix G (other documents). However, if the issuing body has revised or updated the listed version of the standard or document, you may use the new version as guidance if it continues to address the relevant requirements of the Building Regulations.

d. Standards and technical approvals also address aspects of performance or matters that are not covered by the Building Regulations and may recommend higher standards than required by the Building Regulations. Nothing in this guidance precludes you from adopting higher standards.

User requirements

The approved documents provide technical guidance. Users of the approved documents should have adequate knowledge and skills to understand and apply the guidance correctly to the building work being undertaken.

Where you can get further help

If you are not confident that you possess adequate knowledge and skills to apply the guidance correctly or if you do not understand the technical guidance or other information in this approved document or the additional detailed technical references to which it directs you, you should seek further help. Help can be obtained through a number of routes, some of which are listed below.

a. If you are the person undertaking the building work: either from your local authority building control service or from an approved inspector.

b. If you are registered with a competent person scheme: from the scheme operator.

c. If your query is technical: from a specialist or an industry technical body for the relevant subject.

The Building Regulations

The following is a high level summary of the Building Regulations relevant to most types of building work. Where there is any doubt you should consult the full text of the regulations, available at www.legislation.gov.uk.

Building work

Regulation 3 of the Building Regulations defines 'building work'. Building work includes:

a. the erection or extension of a building

b. the provision or extension of a controlled service or fitting

c. the material alteration of a building or a controlled service or fitting.

Regulation 4 states that building work should be carried out in such a way that, when work is complete:

a. For new buildings or work on a building that complied with the applicable requirements of the Building Regulations: the building complies with the applicable requirements of the Building Regulations.

b. For work on an existing building that did not comply with the applicable requirements of the Building Regulations:

   i. the work itself must comply with the applicable requirements of the Building Regulations, and

   ii. the building must be no more unsatisfactory in relation to the requirements than before the work was carried out.

Material change of use

Regulation 5 defines a 'material change of use' in which a building or part of a building that was previously used for one purpose will be used for another.

The Building Regulations set out requirements that must be met before a building can be used for a new purpose. To meet the requirements, the building may need to be altered in some way.

Materials and workmanship

In accordance with regulation 7, building work must be carried out in a workmanlike manner using adequate and proper materials. Guidance on regulation 7(1) is given in Approved Document 7 and guidance on regulation 7(2) is provided in Approved Document B.

Independent third party certification and accreditation

Independent schemes of certification and accreditation of installers can provide confidence that the required level of performance for a system, product, component or structure can be achieved.

Building control bodies may accept certification under such schemes as evidence of compliance with a relevant standard. However, a building control body should establish before the start of the building work that a scheme is adequate for the purposes of the Building Regulations.

Energy efficiency requirements

Part 6 of the Building Regulations imposes additional specific requirements for energy efficiency.

If a building is extended or renovated, the energy efficiency of the existing building or part of it may need to be upgraded.

Notification of work

Most building work and material changes of use must be notified to a building control body unless one of the following applies.

a. It is work that will be self-certified by a registered competent person or certified by a registered third party.

b. It is work exempted from the need to notify by regulation 12(6A) of, or Schedule 4 to, the Building Regulations.

Responsibility for compliance

People who are responsible for building work (e.g. agent, designer, builder or installer) must ensure that the work complies with all applicable requirements of the Building Regulations. The building owner may also be responsible for ensuring that work complies with the Building Regulations. If building work does not comply with the Building Regulations, the building owner may be served with an enforcement notice.

Contents

Section 0: Approved Document B: Fire safety – buildings other than dwellings

Summary

0.1 This approved document has been published in two volumes. Volume 1 deals solely with dwellings, including blocks of flats, while Volume 2 deals with all other types of building covered by the Building Regulations.

Arrangement of sections

0.2 Requirements B1–B5 of Schedule 1 to the Building Regulations are dealt with separately in one or more sections. Each requirement is shown at the start of the relevant sections.

0.3 The provisions in this document have the following aims:

Requirement B1: When there is a fire, ensure both:

a. satisfactory means of sounding an alarm

b. satisfactory means of escape for people.

Requirement B2: Inhibit the spread of fire over internal linings of buildings.

Requirement B3: The building must be built such that all of the following are achieved in the event of a fire:

a. the premature collapse of the building is avoided

b. sufficient fire separation is provided within buildings and between adjoining buildings

c. automatic fire suppression is provided where necessary

d. the unseen spread of fire and smoke in cavities is restricted.

Requirement B4: Restrict both:

a. the potential for fire to spread over external walls and roofs (including compliance with regulations 6(4) and 7(2))

b. the spread of fire from one building to another.

Requirement B5: Ensure both:

a. satisfactory access for the fire service and its appliances

b. facilities in buildings to help firefighters save the lives of people in and around buildings.

Regulation 38: Provide fire safety information to building owners.

0.4 Guidance is given on each aspect separately, though many are closely interlinked. The document should be considered as a whole. The relationship between different requirements and their interdependency should be recognised. Particular attention should be given to the situation where one part of the guidance is not fully followed as this could have a negative effect on other provisions.

Appendices: Information common to more than one requirement of Part B

0.5 Guidance on matters that refer to more than one section of this document can be found in the following appendices.

Appendix A: Key terms

Appendix B: Performance of materials, products and structures

Appendix C: Fire doorsets

Appendix D: Methods of measurement

Appendix E: Sprinklers

Appendix F: Standards referred to

Appendix G: Documents referred to

Management of premises

0.6 The Building Regulations do not impose any requirements on the management of a building, but do assume that it will be properly managed. This includes, for example, keeping protected escape routes virtually 'fire sterile'.

Appropriate fire safety design considers the way in which a building will be managed. Any reliance on an unrealistic or unsustainable management regime cannot be considered to have met the requirements of the regulations.

Once the building is in use, the management regime should be maintained and a suitable risk assessment undertaken for any variation in that regime. Failure to take proper management responsibility may result in the prosecution of an employer, building owner or occupier under legislation such as the Regulatory Reform (Fire Safety) Order 2005.

Property protection

0.7 The Building Regulations are intended to ensure a reasonable standard of life safety in a fire. The protection of property, including the building itself, often requires additional measures. Insurers usually set higher standards before accepting the insurance risk.

Many insurers use the *RISCAuthority Design Guide for the Fire Protection of Buildings* by the Fire Protection Association (FPA) as a basis for providing guidance to the building designer on what they require.

Further information on the protection of property can be obtained from the FPA website: www.thefpa.co.uk.

Inclusive design

0.8 The fire safety aspects of the Building Regulations aim to achieve reasonable standards of health and safety for people in and around buildings.

People, regardless of ability, age or gender, should be able to access buildings and use their facilities. The fire safety measures incorporated into a building should take account of the needs of everyone who may access the building, both as visitors and as people who live or work in it. It is not appropriate, except in exceptional circumstances, to assume that certain groups of people will be excluded from a building because of its use.

The provisions in this approved document are considered to be of a reasonable standard for most buildings. However, some people's specific needs might not be addressed. In some situations, additional measures may be needed to accommodate these needs. This should be done on a case-by-case basis.

Alternative approaches

0.9 The fire safety requirements of the Building Regulations will probably be satisfied by following the relevant guidance in this approved document. However, approved documents provide guidance for some common building situations and there may be alternative methods of complying with the Building Regulations' requirements.

If alternative methods are adopted, the overall level of safety should not be lower than the approved document provides. It is the responsibility of those undertaking the work to demonstrate compliance.

If other standards or guidance documents are adopted, the relevant fire safety recommendations in those publications should be followed in their entirety. However, in some circumstances it may be necessary to use one publication to supplement another. Care must be taken when using supplementary guidance to ensure that an integrated approach is used in any one building.

Guidance documents intended specifically for assessing fire safety in existing buildings often include less onerous provisions than those for new buildings and are therefore unlikely to be appropriate for building work that is controlled by the Building Regulations.

Buildings for industrial and commercial activities that present a special fire hazard, e.g. those that sell fuels, may require additional fire precautions to those in this approved document.

Health care premises

0.10 Health care premises and the patients who use them are diverse. Patients using the premises require different types of care to suit their specific needs. The choice of fire safety strategy depends on both of the following.

a. How a building is designed, furnished, staffed and managed.

b. The needs of the patients.

The Department of Health (DoH) guidance documents on fire precautions in health care buildings, *Firecode*, take account of the particular characteristics of these buildings and should be followed.

Firecode contains managerial and other fire safety provisions that are outside the scope of the Building Regulations.

Unsupervised group homes

0.11 An unsupervised group home for not more than six mental health service users should be regarded as having a purpose group of either of the following.

a. An existing house of one or two storeys for which the means of escape are provided in accordance with DoH HTM 88 should be regarded as a purpose group 1(c) building.

b. A new building may be more appropriately regarded as being in purpose group 2(b).

Shopping complexes

0.12 Although the guidance in this document may be readily applied to individual shops, shopping complexes present different escape problems. The design of units within a shopping complex should be compatible with the fire strategy for the complex as a whole. A suitable approach is given in Annex E of **BS 9999**.

Assembly buildings

0.13 Assembly buildings where a large number of people are present require additional considerations for means of escape; for example, fixed seating may limit the ability of people to escape.

Guidance on fixed seating and other aspects of means of escape in assembly buildings is given in Annex D of **BS 9999**.

For buildings to which the Safety of Sports Grounds Act 1975 applies, the Sports Grounds Safety Authority's *Guide to Safety at Sports Grounds* should also be followed.

Schools

0.14 The design of fire safety in schools is covered by *Building Bulletin 100*, which should be used. *Building Bulletin 100* contains fire safety provisions that are outside the scope of the Building Regulations.

Prisons provided under section 33 of the Prisons Act 1952

0.15 Prisons are exempted from the functional requirements of Parts B1 to B5 of the Building Regulations under section 33 of the Prisons Act 1952. It is usual that prisons should comply with the fire safety requirements of the Building Regulations, except where the requirements are incompatible with safe custody, good order or security.

HM Prison and Probation Service (HMPPS) provides guidance documents on fire precautions in prisons, which take account of the public safety need to secure doors and exits while maintaining life safety objectives.

The HMPPS *Custodial Premises Fire Safety Design Guide* (FSDG) is the design standard for fire safety in prisons, providing structured guidance for those involved in the planning, designing or approval of new or altered buildings.

Further guidance documents on fire safety in prisons are provided by HMPPS. These documents may also be used for other places of lawful detention.

Buildings containing one or more atria

0.16 A building with an atrium that passes through compartment floors may need special fire safety measures. Guidance is given in Annexes B and C of **BS 9999**.

Buildings of special architectural or historic interest

0.17 Where Part B applies to existing buildings, particularly buildings of special architectural or historic interest for which the guidance in this document might prove too restrictive, some variation of the provisions in this document may be appropriate. In such cases, it is appropriate to assess the hazard and risk in the particular case and consider a range of fire safety features in that context.

Fire safety engineering

0.18 Fire safety engineering might provide an alternative approach to fire safety. Fire safety engineering may be the only practical way to achieve a satisfactory standard of fire safety in some complex buildings and in buildings that contain different uses.

Fire safety engineering may also be suitable for solving a specific problem with a design that otherwise follows the provisions in this document.

0.19 **BS 7974** and supporting published documents (PDs) provide a framework for and guidance on the application of fire safety engineering principles to the design of buildings.

Purpose groups

0.20 Building uses are classified within different purpose groups, which represent different levels of hazard (see Table 0.1). A purpose group can apply to a whole building or to a compartment within the building, and should relate to the main use of the building or compartment.

0.21 Where a building or compartment has more than one use, it is appropriate to assign each different use to its own purpose group in the following situations.

a. If the ancillary use is a flat.

b. If both of the following apply.

   i. The building or compartment has an area of more than $280m^2$.

   ii. The ancillary use relates to an area that is more than one-fifth of the total floor area of the building or compartment.

c. In 'shop and commercial' (purpose group 4) buildings or compartments, if the ancillary use is storage and both of the following apply.

   i. The building or compartment has an area of more than $280m^2$.

   ii. The storage area comprises more than one-third of the total floor area of the building or compartment.

0.22 Where there are multiple main uses that are not ancillary to one another (for example, shops with independent offices above), each use should be assigned to a purpose group in its own right. Where there is doubt as to which purpose group is appropriate, the more onerous guidance should be applied.

Table 0.1 Classification of purpose groups

Title	Group	Purpose for which the building or compartment of a building is intended to be used
Volume 1 purpose groups		
Residential (dwellings)	1(a)[1]	Flat.
	1(b)[2]	Dwellinghouse that contains a habitable storey with a floor level a minimum of 4.5m above ground level up to a maximum of 18m.[3]
	1(c)[2][4]	Dwellinghouse that does not contain a habitable storey with a floor level a minimum of 4.5m above ground level.
Volume 2 purpose groups		
Residential (institutional)	2(a)	Hospital, home, school or other similar establishment, where people sleep on the premises. The building may be either of the following: • Living accommodation for, or accommodation for the treatment, care or maintenance of, either: – people suffering from disabilities due to illness or old age or other physical or mental incapacity – people under the age of 5 years. • A place of lawful detention.
Residential (other)	2(b)	Hotel, boarding house, residential college, hall of residence, hostel or any other residential purpose not described above.

Table 0.1 Continued

Title	Group	Purpose for which the building or compartment of a building is intended to be used
Office	3	Offices or premises used for any of the following and their control: • administration • clerical work (including writing, bookkeeping, sorting papers, filing, typing, duplicating, machine calculating, drawing and the editorial preparation of matter for publication, police and fire and rescue service work) • handling money (including banking and building society work) • communications (including postal, telegraph and radio communications) • radio, television, film, audio or video recording • performance (premises not open to the public).
Shop and commercial	4	Shops or premises used for either of the following. • A retail trade or business (including selling food or drink to the public for immediate consumption, retail by auction, self-selection and over-the-counter wholesale trading, the business of lending books or periodicals for gain, the business of a barber or hairdresser, and the rental of storage space to the public). • Premises to which the public are invited either: – to deliver or collect goods in connection with their hire, repair or other treatment – (except in the case of repair of motor vehicles) where the public themselves may carry out such repairs or other treatments.
Assembly and recreation	5	Place of assembly, entertainment or recreation, including any of the following: • bingo halls, broadcasting, recording and film studios open to the public, casinos, dance halls • entertainment, conference, exhibition and leisure centres • funfairs and amusement arcades • museums and art galleries, non-residential clubs, theatres, cinemas, concert halls • educational establishments, dancing schools, gymnasia, swimming pool buildings, riding schools, skating rinks, sports pavilions, sports stadia • law courts • churches and other buildings of worship, crematoria • libraries open to the public, non-residential day centres, clinics, health centres and surgeries • passenger stations and termini for air, rail, road or sea travel • public toilets • zoos and menageries.
Industrial	6	Factories and other premises used for any of the following: • manufacturing, altering, repairing, cleaning, washing, breaking up, adapting or processing any article • generating power • slaughtering livestock.

Table 0.1 Continued

Title	Group	Purpose for which the building or compartment of a building is intended to be used
Storage and other non-residential[4]	7(a)	Either of the following: • place (other than described under 7(b)) for the storage or deposit of goods or materials • any building not within purpose groups 1 to 6.
	7(b)	Car parks designed to admit and accommodate only cars, motorcycles and passenger or light goods vehicles that weigh a maximum of 2500kg gross.

NOTES:

This table only applies to Part B.

See Approved Document B Volume 1 for guidance on dwellings (purpose group 1).

1. Includes live/work units that meet the provisions of Approved Document B Volume 1, paragraph 3.24.
2. Includes any surgeries, consulting rooms, offices or other accommodation that meets all of the following conditions.
   a. A maximum of $50m^2$ in total.
   b. Part of a dwellinghouse.
   c. Used by an occupant of the dwellinghouse in a professional or business capacity.
3. Where very large (over 18m in height or with a 10m deep basement) or unusual dwellinghouses are proposed, some of the guidance for buildings other than dwellings may be needed.
4. All of the following are included in purpose group 1(c).
   a. A detached garage a maximum of $40m^2$ in area.
   b. A detached open carport a maximum of $40m^2$ in area.
   c. A detached building that consists of a garage and open carport, each a maximum of $40m^2$ in area.

Mixed use buildings

0.23 This approved document includes reference to selected guidance for dwellings. For the design of mixed use buildings which include dwellings, Approved Document B Volume 1 should be consulted in addition to the guidance contained in this approved document.

0.24 Where a complex mix of uses exists, the effect that one use may have on another in terms of risk should be considered. It could be necessary to use guidance from both volumes, apply other guidance (such as from HTM 05-02 or *Building Bulletin 100*), and/or apply special measures to reduce the risk.

Requirement B1: Means of warning and escape

These sections deal with the following requirement from Part B of Schedule 1 to the Building Regulations 2010.

Requirement	
Requirement	*Limits on application*
Means of warning and escape	
B1. The building shall be designed and constructed so that there are appropriate provisions for the early warning of fire, and appropriate means of escape in case of fire from the building to a place of safety outside the building capable of being safely and effectively used at all material times.	Requirement B1 does not apply to any prison provided under section 33 of the Prison Act 1952[a] (power to provide prisons, etc.). (a) 1952 c. 52; section 33 was amended by section 100 of the Criminal Justice and Public Order Act 1994 (c. 33) and by S.I. 1963/597.

Intention

In the Secretary of State's view, requirement B1 is met by achieving all of the following.

a. There are sufficient means for giving early warning of fire to people in the building.

b. All people can escape to a place of safety without external assistance.

c. Escape routes are suitably located, sufficient in number and of adequate capacity.

d. Where necessary, escape routes are sufficiently protected from the effects of fire and smoke.

e. Escape routes are adequately lit and exits are suitably signed.

f. There are appropriate provisions to limit the ingress of smoke to the escape routes, or to restrict the spread of fire and remove smoke.

The extent to which any of these measures are necessary is dependent on the use of the building, its size and its height.

Building work and material changes of use subject to requirement B1 include both new and existing buildings.

Section 1: Fire detection and alarm systems

General provisions

1.1 All buildings should have arrangements for detecting fire and raising the alarm. In most buildings, fires are detected by people, either by sight or smell, and therefore often nothing more is needed.

1.2 In some small buildings/premises, the means of raising the alarm may be simple (for example, a shouted warning). In assessing appropriate solutions, warnings need to be heard and understood throughout the premises.

Fire detection and alarm systems

1.3 Other than for some small buildings/premises, an electrically operated fire alarm system should be provided. In some situations, the alarm should be operated by a fire detection system. The detailed specification should be compatible with the fire strategy for the building.

NOTE: The term 'fire alarm system' describes the combination of components for giving an audible and/or other perceptible warning of fire.

NOTE: In this document, the term 'fire detection system' describes any type of automatic sensor network and associated control and indicating equipment. Sensors may be sensitive to smoke, heat, gaseous combustion products or radiation. Automatic sprinkler systems can also be used to operate a fire alarm system.

1.4 In 'residential (institutional)' and 'residential (other)' occupancies (purpose groups 2(a) and 2(b)), automatic fire detection and alarms should be provided.

1.5 Automatic fire detection and alarm systems should be provided in non-residential occupancies where a fire could break out in an unoccupied part of the premises (e.g. a storage area or a part of the building that is not visited on a regular basis) and prejudice the means of escape from occupied part(s) of the premises.

1.6 Automatic fire detection will also be necessary where fire protection systems, such as pressure differential systems or door releases, need to operate automatically.

1.7 Every building design should be assessed individually. General guidance on the category of fire detection system that *may* need to be provided within a building can be found in Table A1 of **BS 5839-1**.

1.8 Where an electrically operated fire detection and alarm system is provided, it should comply with **BS 5839-1**.

1.9 **BS 5839-1** specifies three categories of system.

a. Category L – for the protection of life.

b. Category M – manual fire detection and alarm systems.

c. Category P – for property protection.

Category L systems are divided into the following.

L1 – systems installed throughout the protected building.

L2 – systems installed only in defined parts of the protected building (a category L2 system will normally include the coverage required of a category L3 system).

L3 – systems designed to warn of fire at an early enough stage to enable all occupants, other than possibly those in the room where the fire started, to escape safely before the escape routes become impassable because of fire, smoke or toxic gases.

L4 – systems installed within those parts of the escape routes that comprise circulation areas and circulation spaces, such as corridors and stairs.

L5 – systems in which the protected area(s) and/or the location of detectors are designed to satisfy a specific fire safety objective (other than that of a category L1, L2, L3 or L4 system).

Type P systems are divided into the following.

P1 – systems installed throughout the protected building.

P2 – systems installed only in defined parts of the protected building.

1.10 Electrical alarm system call points should comply with either of the following.

a. **BS 5839-2**.

b. **BS EN 54-11** Type A (direct operation).

Call points should be installed in accordance with **BS 5839-1**.

Type B (indirect operation) call points of **BS EN 54-11** should only be used with the approval of the building control body.

1.11 A voice alarm system complying with **BS 5839-8**, and giving a fire warning different from other signals in general use, may be considered if either of the following applies.

a. People might not respond quickly to a fire warning.

b. People are unfamiliar with the fire warning arrangements.

1.12 In premises where lots of members of the public are present, an initial general alarm may be undesirable. Any fire alarm system that first alerts staff should comply with **BS 5839-1**.

1.13 Where the escape strategy is based on simultaneous evacuation, actuation of the fire alarm system should give warning from all fire alarm sounders. Where phased evacuation is planned, a staged alarm system is appropriate. See paragraph 3.21.

1.14 **BS 9999** provides guidance for fire detection and alarm systems in buildings containing atria.

Warnings for people with impaired hearing

1.15 Clause 18 of **BS 5839-1** gives detailed guidance on the design and selection of fire alarm warnings for people with impaired hearing. In buildings or part of a building where people may be in relative isolation, a visual and audible fire alarm may be the most appropriate solution. In buildings where the population is managed, a vibrating personal paging system may be more appropriate.

Design and installation of systems

1.16 Fire detection and alarm systems must be properly designed, installed and maintained. A design, installation and commissioning certificate should be provided for fire detection and alarm systems. Third party certification schemes for fire protection products and related services are an effective means of providing assurances of quality, reliability and safety.

Interface between fire detection and alarm systems and other systems

1.17 Fire detection and alarm systems sometimes trigger other systems. The interface between systems must be reliable. Particular care should be taken if the interface is facilitated via another system. Where any part of **BS 7273** applies to the triggering of other systems, the recommendations of that part of **BS 7273** should be followed.

Section 2: Design for horizontal escape

Introduction

2.1 Means of escape should be provided from any point on a storey to the storey exit, for all types of building. The general principle is that any person confronted by a fire within a building can turn away from it and escape safely.

2.2 For small shop, office, industrial, storage and other similar premises, the guidance on small premises (see section 4) may be followed instead of the provisions in this section, if they meet both of the following conditions.

a. No storey has an area more than $280m^2$.

b. There is a maximum of two storeys plus a basement storey.

Escape route design

Number of escape routes and exits

2.3 The number of escape routes and exits that should be provided depends on both of the following.

a. The number of occupants in the room, tier or storey.

b. The limits on travel distance to the nearest exit given in Table 2.1 (which apply only to the nearest exit; other exits may be further away).

2.4 In multi-storey buildings, if more than one stair is needed for vertical escape, every part of each storey should have access to more than one stair. An area may be in a dead end provided the alternative stair is accessible.

2.5 In mixed use buildings, separate means of escape should be provided from any storeys or parts of storeys used for the 'residential' or 'assembly and recreation' purpose groups (purpose groups 1, 2 and 5).

Single escape routes and exits

2.6 A single escape route is acceptable for either of the following.

a. Parts of a floor from which a storey exit can be reached within the limit for travel distance in one direction shown in Table 2.1 (see also paragraph 2.8), provided the following apply.

i. For places of assembly and bars, no one room in this situation has more than 60 people.

ii. For 'residential (institutional)' buildings (purpose group 2(a)), no one room in this situation has more than 30 people. Occupant number calculations are described in Appendix D.

b. A storey with no more than 60 people, where the limits on travel distance in one direction only are satisfied (see Table 2.1).

2.7 In many cases, the beginning of a route will not have an alternative escape route (for example, a single exit from a room into a corridor where escape is possible in two directions). This is acceptable if both of the following apply.

a. The travel distance to the nearest storey exit is within the limits for routes where escape is possible in more than one direction (Table 2.1).

b. The travel distance for the 'one direction only' section of the route does not exceed the limit for travel distance where there is no alternative escape route (Table 2.1).

Diagram 2.1 shows how to measure travel distances from a dead end in an open storey layout.

See para 2.7

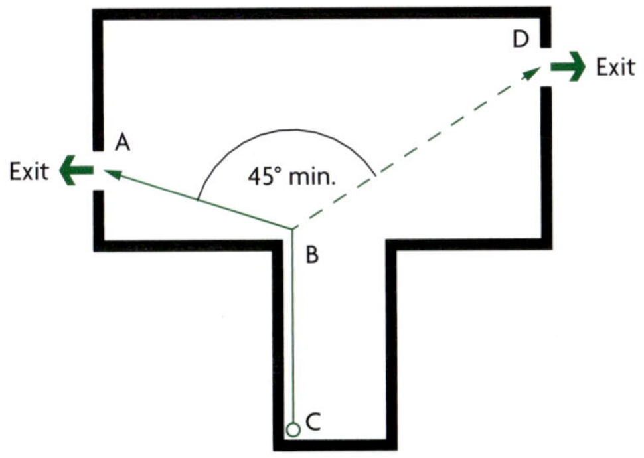

Travel distance in a dead-end condition should meet all of the following.

a. Angle ABD should be at least 45 degrees.

b. Distance CBA or CBD (whichever is less) should be no more than the maximum travel distance given for alternative escape routes.

c. Distance CB should be no more than the maximum travel distance where there are no alternative escape routes.

Diagram 2.1 Travel distance in dead-end condition

Table 2.1 Limitations on travel distance

Purpose group	Use of the premises or part of the premises		Maximum travel distance[1] where travel is possible in: One direction only (m)	More than one direction (m)
2(a)	Residential (institutional)		9	18
2(b)	Residential (other):			
	a. in bedrooms[2]		9	18
	b. in bedroom corridors		9	35
	c. elsewhere		18	35
3	Office		18	45
4	Shop and commercial		18	45
5	Assembly and recreation:			
	a. buildings primarily for disabled people		9	18
	b. areas with seating in rows		15	32
	c. elsewhere		18	45
6	Industrial[3]	Normal hazard	25	45
		Higher hazard	12	25
7	Storage and other non-residential[3]	Normal hazard	25	45
		Higher hazard	12	25
2–7	Place of special fire hazard[4]		9[5]	18[5]
2–7	Plant room or roof-top plant:			
	a. distance within the room		9	35
	b. escape route not in open air (overall travel distance)		18	45
	c. escape route in open air (overall travel distance)		60	100

NOTES:

1. If the internal layout of partitions, fittings, etc. is not known, direct distances, rather than travel distances, should be assessed. The direct distance should be assumed to be two-thirds of the actual travel distance.
2. Maximum part of travel distance within the room. This limit applies within the bedroom and any associated dressing room, bathroom or sitting room, etc. The distance is measured to the door to the protected corridor that serves the room or suite. Sub-item (b) applies from that point along the bedroom corridor to a storey exit.
3. In industrial and storage buildings, the appropriate travel distance depends on the level of fire hazard associated with the processes and materials being used.

   Higher hazard includes manufacturing, processing or storage of significant amounts of hazardous goods or materials, including any of the following.

   - Any compressed, liquefied or dissolved gas.
   - Any substance that becomes dangerous by interaction with either air or water.
   - Any liquid substance with a flash point below 65°C, including whisky or other alcoholic liquor.
   - Any corrosive substance.
   - Any oxidising agent.
   - Any substance liable to spontaneous combustion.
   - Any substance that changes or decomposes readily, giving out heat when doing so.
   - Any solid substance with a flash point less than 120°C.
   - Any substance that is likely to spread fire by flowing from one part of a building to another.
4. Places of special fire hazard are listed in the definitions in Appendix A.
5. Maximum part of travel distance within the room/area. Travel distance outside the room/area should comply with the limits for the purpose group of the building or part.

Access control measures

2.8 Measures to restrict access to the building (or parts of it) should not adversely affect fire safety provisions. It may be reasonable to close some escape routes outside normal business hours, but measures should remain to safely evacuate people left inside the building (see paragraph 5.6).

Number of occupants and exits

2.9 The building design should be based on the number of occupants. If the number is not known, use the appropriate floor space factors (Appendix D).

Table 2.2 gives the minimum number of escape routes and exits from a room or storey for different numbers of occupants. This number is likely to be increased by the need to observe travel distances and other practical considerations.

The width of escape routes and exits is given in paragraph 2.18.

Table 2.2 Minimum number of escape routes and exits from a room, tier or storey

Maximum number of people	Minimum number of escape routes/exits
60	1
600	2
More than 600	3

Alternative escape routes

2.10 Alternative escape routes should satisfy one of the following criteria.

a. They are in directions 45 degrees or more apart (Diagram 2.2).

b. They are in directions less than 45 degrees apart, but separated from each other by fire resisting construction.

See para 2.10

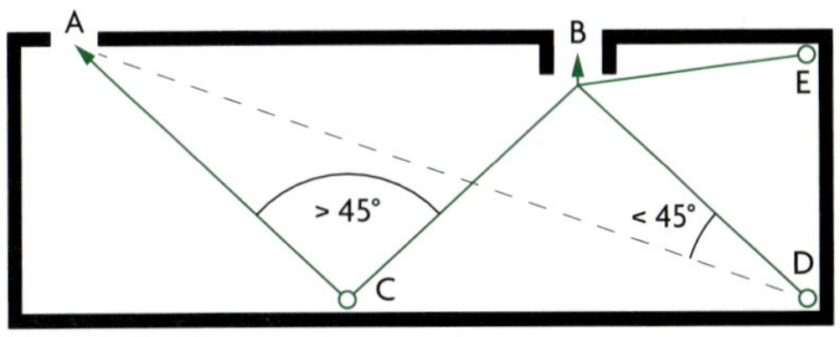

Alternative escape routes are available from C because angle ACB is 45 degrees or more and therefore distance CA or CB (whichever is the less) should be no more than the maximum travel distance given for alternative escape routes.

Alternative escape routes are not available from D because angle ADB is less than 45 degrees (therefore see Diagram 2.1).

There is also no alternative escape route from E.

Diagram 2.2 Alternative escape routes

Inner rooms

2.11 An inner room is at risk if a fire starts in the access room (Diagram 2.3). Such an arrangement should only be accepted if all of the following conditions are satisfied.

a. The occupant number of the inner room *does not* exceed:

   i. 30 people for 'residential (institutional)' buildings (purpose group 2(a))

   ii. 60 people for other purpose groups.

b. The inner room is not a bedroom.

c. The inner room is entered directly from the access room (but not via a corridor).

d. The escape route from the inner room does not pass through more than one access room.

e. The travel distance from any point in the inner room to the exits from the access room does not exceed the distances in Table 2.1.

f. The access room meets both of the following conditions.

   i. It is not a place of special fire hazard.

   ii. It is in the control of the same occupier.

g. One of the following arrangements is made.

   i. The enclosures (walls or partitions) of the inner room stop a minimum of 500mm below the ceiling.

   ii. The door or walls of the inner room contain a vision panel (minimum $0.1m^2$), so people can see if a fire starts in the access room.

   iii. The access room is fitted with an automatic fire detection and alarm system to warn occupants of the inner room if a fire starts in the access room.

See para 2.11

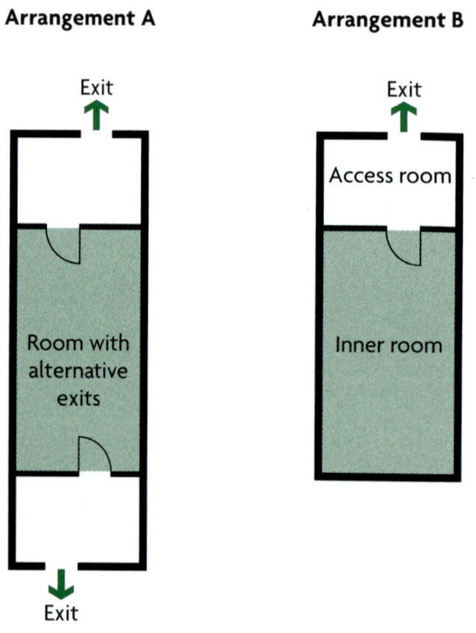

NOTES:

Arrangement A Needs no special provision.

Arrangement B Should observe the inner room provisions in paragraph 2.11.

Diagram 2.3 Inner room and access room

Planning of exits in a central core

2.12 Where a central core has more than one exit, storey exits should be remote from one another and no two exits should be approached from the same lift hall, common lobby or undivided corridor (Diagram 2.4).

Open spatial planning

2.13 Escape routes should not be within 4.5m of openings between floors, such as for an escalator, unless either of the following applies.

a. The direction of travel is away from the opening.

b. An alternative escape route does not pass within 4.5m of the open connection (Diagram 2.5).

See para 2.12

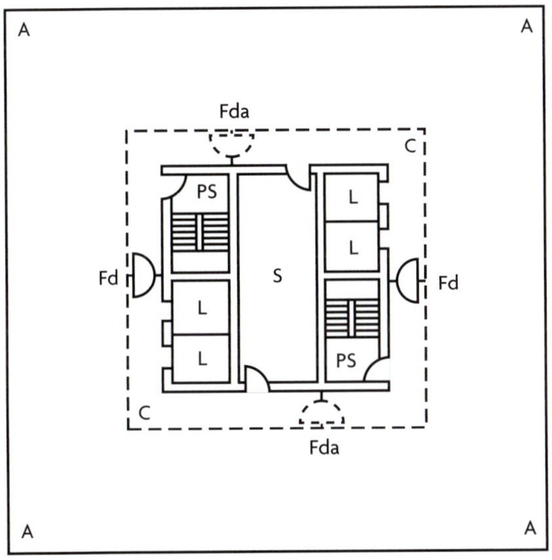

NOTE: The doors at both ends of the area marked 'S' should be self-closing fire doorsets unless the area is sub-divided such that any fire in that area will not be able to prejudice both sections of corridor at the same time. If that area is a lift lobby, doors should be provided as shown in Figure 9 in **BS 9999**.

L	Lift
S	Services, toilets, etc.
Fd	Self-closing E 20 S_a fire doorsets
Fda	Possible alternative position for fire doorset
C	Corridor off which accommodation opens
PS	Protected stairway
A	Accommodation (e.g. office space)

Diagram 2.4 Exits in a central core

See para 2.13

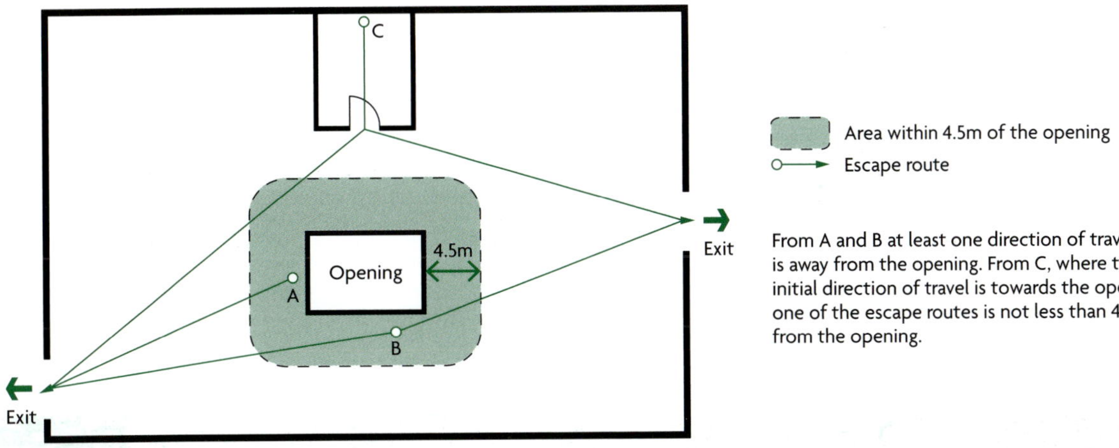

From A and B at least one direction of travel is away from the opening. From C, where the initial direction of travel is towards the opening, one of the escape routes is not less than 4.5m from the opening.

Diagram 2.5 Open connections

Access to storey exits

2.14 Where a storey has more than one escape stair, it should be planned so that it is not necessary to pass through one stair to reach another. However, it would be acceptable to pass through one stair's protected lobby to reach another stair.

Separation of circulation routes from protected stairways

2.15 Where they serve protected stairways that are part of primary circulation routes, self-closing fire doors should be fitted with an automatic release mechanism, to avoid them being rendered ineffective by misuse. Otherwise, the stair (and any associated exit passageway) *should not* form part of the primary circulation route between different parts of the building at the same level.

Storeys divided into different uses

2.16 If a storey contains areas for consuming food and/or drink, and where that is not the main use of the building, then both of the following apply.

a. A minimum of two escape routes should be provided from each area, except from inner rooms that meet the conditions in paragraph 2.11.

b. Those escape routes should lead directly to a storey exit without entering a kitchen or similar area of high fire hazard.

Storeys divided into different occupancies

2.17 Where a storey is divided into areas of occupancy under separate ownership or tenancy, then both of the following apply.

a. The means of escape from each occupancy should not pass through any other occupancy.

b. If a common corridor or circulation space is on the escape route, one of the following should apply.

   i. It should be a protected corridor.

   ii. A suitable automatic fire detection and alarm system should be installed throughout the storey.

Width of escape routes and exits

2.18 The width of escape routes and exits should meet the provisions in Table 2.3, as well as the guidance in Approved Document M.

2.19 If the maximum number of people likely to use the escape route and exit is not known, it should be calculated using the occupant number guidance in Appendix D.

2.20 Guidance on the spacing of fixed seating for auditoria is given in Annex D of **BS 9999**.

Table 2.3 Widths of escape routes and exits

Maximum number of people	Minimum width (mm)[1][2][3]
60	750[4]
110	850
220	1050
More than 220	5 per person[5]

NOTES:

1. See Appendix D for methods of measurement.
2. Widths may need to be increased to meet guidance in Approved Document M.
3. Widths less than 1050mm should not be interpolated.
4. May be reduced to 530mm for gangways between fixed storage racking, other than in public areas of 'shop and commercial' (purpose group 4) buildings.
5. 5mm/person does not apply to an opening serving fewer than 220 people.

Calculating exit capacity

2.21 Where multiple storey exits are available, fire might prevent one from being used. Remaining exits need to be wide enough for all occupants, so when using Table 2.3, the largest exit should be discounted.

Stairs should be at least as wide as any storey exit leading onto them. While some stairs are not subject to discounting (paragraphs 3.14 and 3.15), because the stairs will be available for other floors, the storey exits onto them are.

2.22 To calculate how many people two or more available exits (after discounting) can accommodate, add together the maximum numbers of people that each exit width can accommodate.

For example, three exits each 850mm wide accommodate 3×110 = 330 people.

2.23 If a ground floor storey exit and a stair share a final exit (via a ground floor lobby), then the final exit should be wide enough to evacuate people at a maximum flow rate equal to or greater than from the storey exit and stair combined (Diagram 2.6).

See para 2.23

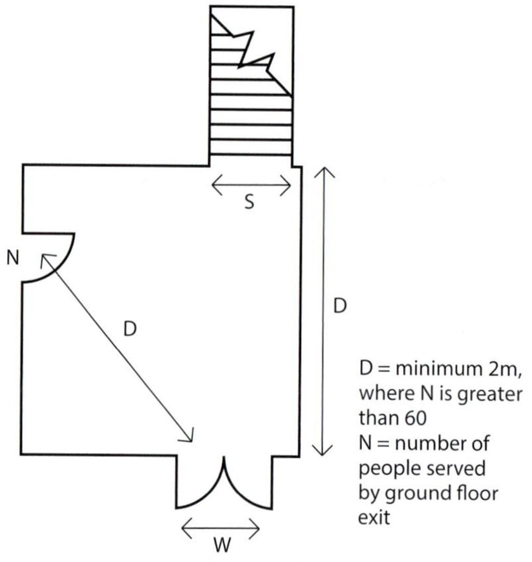

Diagram 2.6 Merging flows at final exit

This can be calculated using the following formula:

W = ((N/2.5) + (60S))/80

where:

W is the width of final exit in metres

N is the number of people served by ground floor storey exit

S is the stair width in metres.

If the number of people (N) entering the lobby from the ground storey is more than 60, then the distance from the foot of the stair or the storey exit to the final exit should be a minimum of 2m (see Diagram 2.6).

If that minimum distance cannot be achieved, the width of the final exit (W) should be at least the width of the stair plus the width of the storey exit.

Worked example

A ground floor storey exit serving 250 people shares a common final exit with a 1.2m wide stair.

Required final exit width = ((250/2.5) + (1.2×60))/80 = 2.150m

Protected corridors

2.24 A corridor serving as part of the means of escape in any of the following circumstances should be a protected corridor.

a. Every corridor that serves bedrooms.

b. Every dead-end corridor (excluding recesses and extensions a maximum of 2m deep, as shown in Diagrams 2.7 and 2.8).

c. Any corridor shared by two or more occupancies (paragraph 2.17).

See para 2.24

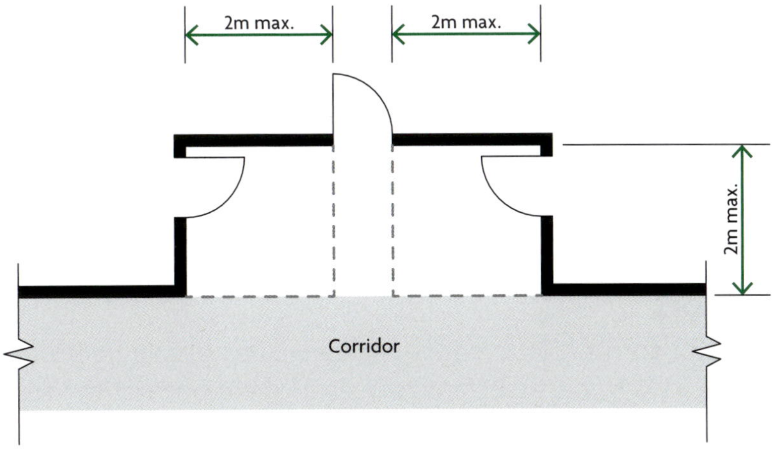

Diagram 2.7 Recesses off corridors

See para 2.24

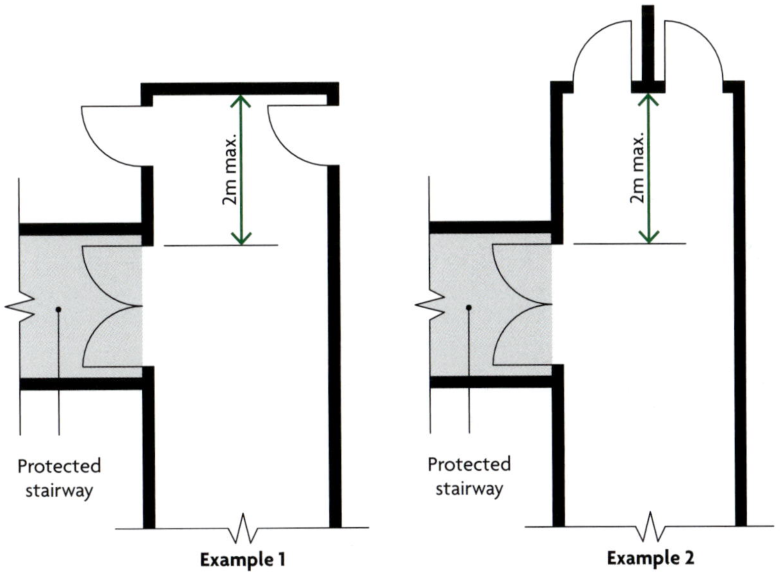

Diagram 2.8 Extension of corridor beyond a protected stairway

Enclosure of corridors that are not protected corridors

2.25 If a corridor is used for a means of escape but is not a protected corridor, even though the enclosing partitions may have no fire resistance, both of the following should be met to inhibit the spread of smoke.

a. Partitions should continue to the soffit of the structural floor above, or to a suspended ceiling.

b. Openings into rooms from the corridor should be fitted with doors, which do not need to be fire doorsets.

Open planning will not inhibit the spread of smoke, but occupants can become aware of a fire quickly.

Division of corridors

2.26 A corridor providing access to alternative escape routes *should be divided by fire doorsets fitted with a self-closing device* (and associated screens) where both of the following apply.

a. It is more than 12m long.

b. It connects two or more storey exits.

The fire doorsets (including any screens) should be approximately mid-way between the two storey exits. They should safeguard the route from smoke, while considering the layout of the corridor and any adjacent fire risks.

2.27 For buildings other than dwellings (purpose groups 2 to 7): if a cavity exists above the enclosures to a corridor as described above (because the enclosures are not carried to full storey height or the underside of the roof covering at the top storey), the potential for smoke to bypass the enclosure should be restricted by one of the following methods.

a. Method 1 – Fitting cavity barriers on the line of the enclosure(s) to and across the corridor (Diagram 2.9).

b. Method 2 – Dividing the storey using fire resisting construction that passes through the line of the division of the corridor (Diagram 2.9). Any cavity above this division should be fitted with cavity barriers on the line of division of the storey and the corridor.

c. Method 3 – Enclosing the cavity on the lower side by a fire resisting ceiling that extends throughout the building, compartment or separated part.

Any door that could provide a path for smoke to bypass the division should be fitted with a self-closing device (but need not necessarily be fire resisting).

See para 2.27

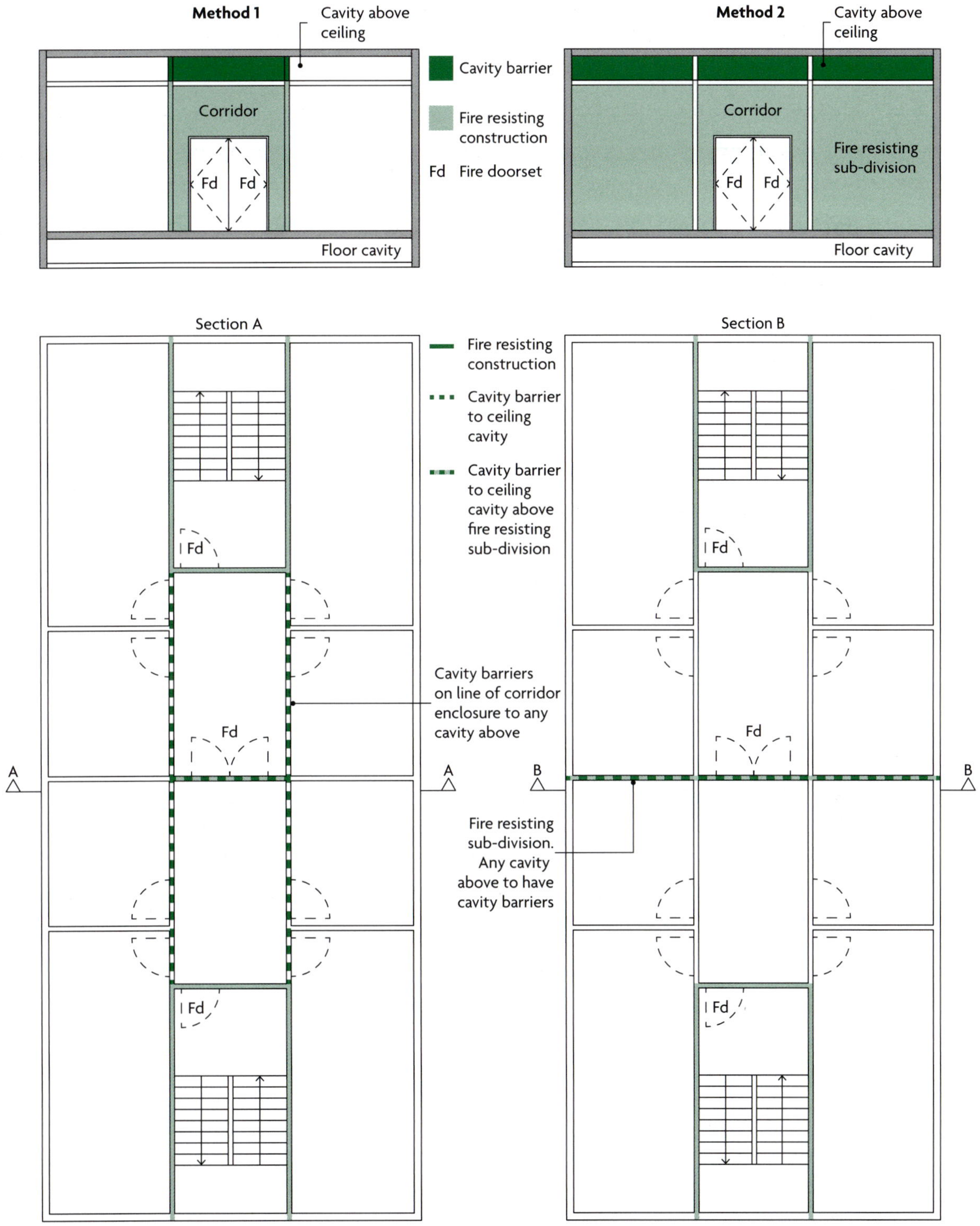

NOTES:

For all methods, where the corridor is a protected escape route, cavity barriers may also be required in any floor cavity beneath the corridor enclosure (see paragraph 9.5).

The sub-division should be carried to full storey height and includes sub-division of the corridor. A cavity barrier may be used in any ceiling cavity over the sub-division.

Diagram 2.9 Division of corridors

2.28 Where dead ends of corridors exceeding 4.5m long provide access to a point from which alternative escape routes are available, they should be separated by self-closing fire doorsets (together with any associated screens) from any part of the corridor that either:

a. Provides two directions of escape (Diagram 2.10a)

b. Continues past one storey exit to another (Diagram 2.10b).

Alternatively, the stairs and corridors may be protected by a pressurisation system complying with **BS EN 12101-6**.

See para 2.28

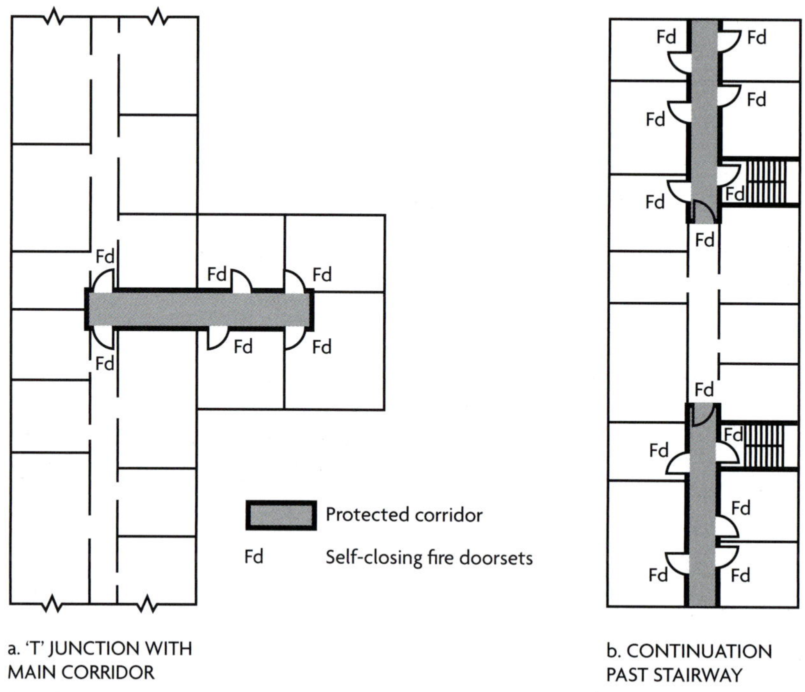

Diagram 2.10 Dead-end corridors

Cavity barriers

2.29 Additional measures to safeguard means of escape from smoke are given in Section 10.

External escape routes

2.30 Where an external escape route is beside an external wall of the building, the external wall should be of fire resisting construction in both of the following zones.

a. Within 1800mm of the escape route.

b. Up to 1100mm above the surface of the escape route.

This *does not* apply to external escape stairs (see paragraph 3.32).

Escape over flat roofs

2.31 Where a storey or part of a building has multiple escape routes available, one may be over a flat roof if it does not serve a 'residential (institutional)' (purpose group 2(a)) building, or part of a building intended for use by members of the public.

2.32 Where an escape route over a flat roof is provided, the roof should comply with all of the following.

a. It should be part of the same building from which escape is being made.

b. The route across the roof should lead to a storey exit or external escape route.

c. The part of the roof forming the escape route and its supporting structure, together with any opening within 3m of the escape route, should be fire resisting (minimum REI 30).

d. The route should be clearly defined and guarded by walls/or protective barriers to protect from falling.

Residential care homes

General provisions

2.33 The choice of fire safety strategy depends on the way a building is designed, furnished, staffed and managed, and on the level of dependency of the residents.

2.34 In care homes for the elderly, some or all residents are likely to need help to evacuate. Buildings should generally be designed for progressive horizontal evacuation (PHE) in accordance with paragraphs 2.35 to 2.46.

For other care home types, the most appropriate of either a PHE or simultaneous evacuation strategy should be identified. The approach adopted in the design of a building must be recorded and communicated to the building management team, who can adopt procedures compatible with the building design.

Planning for progressive horizontal evacuation

2.35 The guidance below on PHE is for care homes where the provisions of the *Firecode* documents *do not* apply (see paragraph 0.10).

PHE requires areas used for the care of residents to be divided into protected areas by compartment walls and compartment floors. Protected areas provide a place of relative safety, from which further evacuation can be made if necessary.

2.36 Each storey used for the care of residents should be divided by compartment walls into at least three protected areas. All floors should be compartment floors.

2.37 Every protected area should have a minimum of two exits to adjoining protected areas. Maximum travel distances within a protected area should be both of the following.

a. To the exit to the adjoining protected area: as shown in Table 2.1.

b. From any point to a storey exit or a final exit: 64m.

2.38 A fire in one protected area should not prevent occupants of other areas from reaching a final exit (Diagram 2.11). Escape routes *should not* pass through ancillary accommodation listed in paragraph 2.44.

2.39 The number of residents' beds in protected areas should be based on an assessment of both of the following.

a. The number of staff likely to be available.

b. The level of assistance that residents may require.

The maximum number of residents' beds in one protected area *should not* exceed 10, but may need to be lower depending on the assessment.

2.40 A protected area used for horizontal evacuation from an adjoining protected area should have a floor area able to accommodate its own occupants plus those from the largest adjoining protected area.

See para 2.38

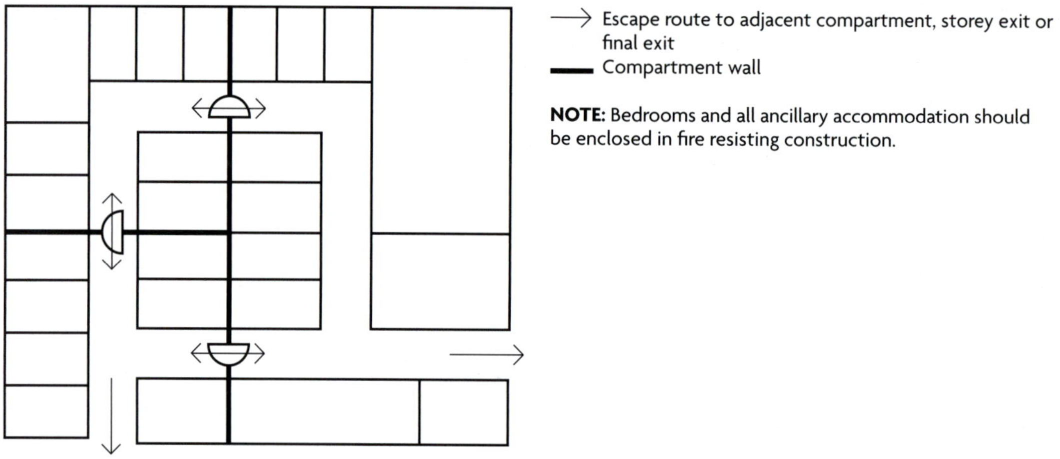

Diagram 2.11 Progressive horizontal evacuation in care homes

Fire detection and alarm

2.41 A fire detection and alarm system should be provided to L1 standard in accordance with **BS 5839-1**.

Bedrooms

2.42 Each bedroom in a care home should be enclosed in fire resisting construction (minimum REI 30) with fire resisting doors (minimum E 30). Every corridor serving bedrooms should be a protected corridor (see paragraph 2.24).

2.43 Bedrooms should not contain more than one single or double bed.

Ancillary accommodation

2.44 Ancillary accommodation such as all of the following should be enclosed by fire resisting construction (minimum REI 30).

a. Chemical stores.

b. Cleaners' rooms.

c. Clothes storage.

d. Day rooms.

e. Smoking rooms.

f. Disposal rooms.

g. Plant rooms.

h. Linen stores.

i. Kitchens.

j. Laundry rooms.

k. Staff changing and locker rooms.

l. Store rooms.

Door closing devices

2.45 If doors fitted with a self-closing device could present an obstacle to residents, the following hardware in accordance with **BS EN 1155** is appropriate.

a. Bedrooms: free-swing door closers.

b. Circulation spaces: hold-open devices.

Sprinkler systems

2.46 When a sprinkler system is provided in accordance with Appendix E, the following variations to the guidance given in paragraphs 2.35 to 2.45 are acceptable.

a. Fire doorsets to bedrooms do not need to be fitted with self-closing devices.

b. Protected areas may contain more than 10 beds.

c. Bedrooms may contain more than one bed.

If any of the variations are made, the management procedures should take account of the larger number of residents that may need assistance, and the need to manually close bedroom doors during sleeping hours.

Section 3: Design for vertical escape

Introduction

3.1 The limits on horizontal travel escape distances mean most people should be able to independently reach a protected escape route or final exit. The following guidance also includes measures for people who are unable to use stairs without help.

In larger buildings, some escape stairs may need to serve as firefighting stairs, and Section 17 will also apply.

Number of escape stairs

Mixed use buildings

3.2 If a building contains storeys, or parts of storeys, in different purpose groups, it is necessary to consider providing either of the following.

a. Separate escape routes from the areas of different use.

b. Other effective means to protect common escape routes.

Single escape stairs

3.3 A single escape stair may serve a building (or part of a building) in the following situations.

a. When independent escape routes from areas in different purpose groups are not necessary (see paragraph 3.2).

b. From a basement that is allowed to have a single escape route in accordance with paragraph 2.6b and Table 2.1.

c. In small premises, provided it meets the conditions in paragraph 4.2.

d. From a building that meets both of the following conditions.

   i. It has no storey with a floor level more than 11m above ground level.

   ii. It is allowed to have only a single escape route in every storey in accordance with paragraph 2.6b and Table 2.1.

e. An office building with a maximum of five storeys above the ground storey where both of the following apply.

   i. The travel distance from every point in each storey does not exceed the distances given in Table 2.1 for escape in one direction only.

   ii. Every storey with a floor level more than 11m above ground level has an alternative means of escape.

f. A factory comprising no more than either of the following.

   i. For low risk buildings, two storeys above the ground storey.

   ii. For normal risk buildings, one storey above the ground storey, provided the travel distance from every point on each storey does not exceed the distances given in Table 2.1 for escape in one direction only.

g. Process plant buildings with a maximum of 10 people.

Provision of refuges

3.4 Refuges form part of the management plan and offer relatively safe areas for people to wait for a short period only. Refuges should meet the following conditions.

a. Refuges should be provided on every storey (except ones consisting only of plant rooms) of each protected stairway providing an exit from that storey.

b. Refuges do not need to be located within the stair enclosure, but should enable direct access to the stair.

c. The number of refuge spaces does not need to equal the number of wheelchair users who may be in the building. A single refuge may be occupied by more than one person during the evacuation procedure.

3.5 The following are both examples of satisfactory refuges.

a. An enclosure such as a compartment (Diagram 3.1), protected lobby, protected corridor or protected stairway (Diagram 3.2).

b. An area in the open air, such as a flat roof, balcony, podium or similar place, that meets both of the following.

   i. It is protected (or remote) from any fire risk.

   ii. It has its own means of escape.

3.6 Refuges should be a minimum of 900mm × 1400mm in size and accessible by someone in a wheelchair. Where sited in a protected stairway, protected lobby or protected corridor, they should not reduce the width of the escape route or obstruct the flow of people escaping.

3.7 Refuges should be provided with an emergency voice communication (EVC) system complying with **BS 5839-9**. It should consist of Type B outstations communicating with a master station in the building control room (if one exists) or next to the fire detection and alarm panel. In some buildings, wireless technology may be more appropriate.

3.8 Refuges and evacuation lifts should be clearly identified. In protected lobbies and protected stairways there should be a blue mandatory sign worded 'Refuge – keep clear' in addition to fire safety signs.

3.9 Paragraph 5.32 gives guidance on using lifts, including evacuation lifts, during a fire.

See para 3.5

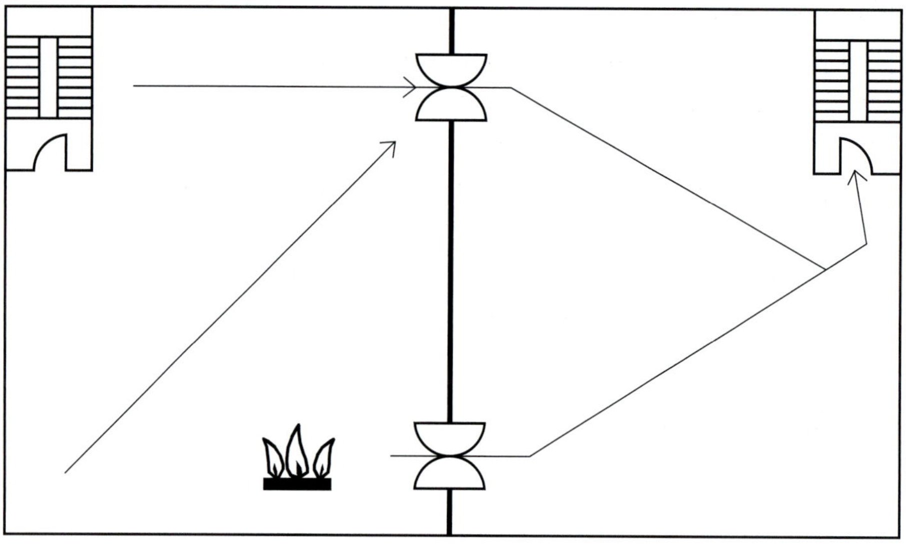

Storey divided into two refuges by compartment wall (stairways not provided with wheelchair space).

NOTE: People occupying the left-hand compartment would not reach a refuge until they had entered the right-hand compartment. Two fire doorsets in the partition are necessary in case access to one of the doorsets is blocked by fire.

Diagram 3.1 Refuge formed by compartmentation

See para 3.5

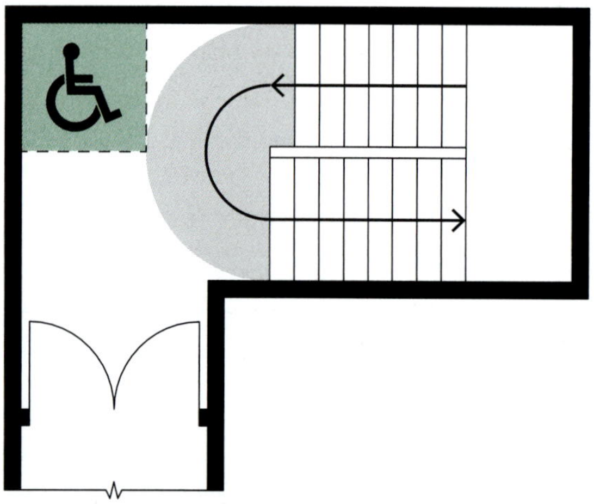

Provision where access to the refuge is counter to the access flow within the stairway.

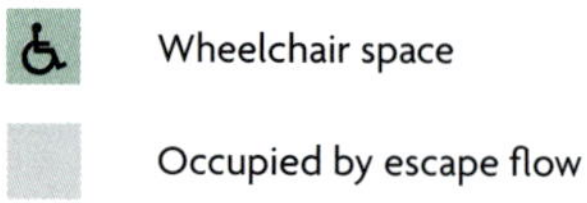

Diagram 3.2 Refuge formed in a protected stairway

Width of escape stairs

3.10 The width of escape stairs should meet all of the following conditions.

a. It should be at least as wide as any exits giving access to the stairs.

b. It should be no less than the minimum widths given in Table 3.1.

c. It should not reduce at any point on the way to a final exit.

d. It should not exceed 1400mm in stairs taller than 30m, unless a central handrail is provided. When a central handrail is provided, the stair width on each side of it should be considered separately when assessing stair capacity.

3.11 Approved Document K requires stairs more than 2000mm wide in public buildings to have a central handrail.

3.12 If an exit route from a stair is also the escape route from the ground storey and/or basement storey, the width of the exit route may need to be increased (see paragraph 2.23).

Table 3.1 Minimum widths of escape stairs

Situation of stair	Maximum number of people served[1]	Minimum stair width (mm)
1a. In a 'residential (institutional)' building (unless the stair will only be used by staff)	150	1000[2]
1b. In an 'assembly and recreation' building and serving an area used for assembly purposes (unless the area is less than $100m^2$)	220	1100
1c. In any other building and serving an area with an occupancy of more than 50	Over 220	See note 3
2. Any stair not described above	50	800[4]

NOTES:

1. Assessed as likely to use the stair in a fire emergency.
2. Section 6 of **BS 9999** recommends that firefighting stairs should be at least 1100mm wide.
3. See Table 3.2 for the size of stairs for simultaneous evacuation, and Table 3.3 for phased evacuation.
4. To comply with the guidance in Approved Document M on minimum widths for areas accessible to disabled people, this may need to be increased to 1000mm.

Calculation of minimum stair width

3.13 The width depends on the number of stairs provided and the escape strategy (simultaneous or phased evacuation). If the maximum number of people needing to use escape stairs is unknown, calculate it using the floor space factors in Appendix D.

Discounting of stairs

3.14 Regardless of escape strategy, where two or more stairs are provided, it should be assumed that one might not be available during a fire. Each stair should be discounted in turn to ensure the capacity of the remaining stairs is adequate. This applies to buildings with or without a sprinkler system.

3.15 Paragraph 3.14 does not apply if either of the following applies.

a. Escape stairs are protected by a smoke control system designed in accordance with **BS EN 12101-6**.

b. Escape stairs are approached on each storey (except the top storey) through a protected lobby.

Despite these exceptions, at least one storey exit still needs to be discounted (paragraph 2.21).

Paragraph 3.34 identifies cases where stairs need lobby protection.

Simultaneous evacuation

3.16 The width of escape stairs should take account of the number of people using them while evacuating all storeys at the same time. The following stairs should be designed to allow simultaneous evacuation.

a. All stairs serving basements.

b. All stairs serving buildings with open spatial planning.

c. All stairs serving 'residential (other)' (purpose group 2(b)) or 'assembly and recreation' (purpose group 5) buildings.

Annexes B and C of **BS 9999** include designs based on simultaneous evacuation.

3.17 The capacity of stairs of widths from 1000mm to 1800mm is given in Table 3.2.

Table 3.2 Capacity of stairs for basements and for simultaneous evacuation of the building

No. of floors served	Maximum number of people served by a stair of width:								
	1000mm	1100mm	1200mm	1300mm	1400mm	1500mm	1600mm	1700mm	1800mm
1	150	220	240	260	280	300	320	340	360
2	190	260	285	310	335	360	385	410	435
3	230	300	330	360	390	420	450	480	510
4	270	340	375	410	445	480	515	550	585
5	310	380	420	460	500	540	580	620	660
6	350	420	465	510	555	600	645	690	735
7	390	460	510	560	610	660	710	760	810
8	430	500	555	610	665	720	775	830	885
9	470	540	600	660	720	780	840	900	960
10	510	580	645	710	775	840	905	970	1035

NOTES:

1. The capacity of stairs that serve more than 10 storeys may be obtained by using linear extrapolation.
2. The capacity of stairs not less than 1100mm wide may also be obtained by using the formulas in paragraph 3.18.
3. Unless a central handrail is provided, stairs with a rise of more than 30m should be a maximum width of 1400mm (see paragraph 3.10).
4. Stairs wider than 2000mm should have a central handrail (see paragraph 3.11).

3.18 As an alternative to Table 3.2, the capacity of stairs 1100mm wide or wider can be found using either of the following formulas:

a. $P = 200W + 50 (W - 0.3)(N - 1)$

b. $W = P + 15N - 15 / 150 + 50N$

where:

P is the number of people that can be served

W is the width of the stair, in metres

N is the number of storeys served.

Separate calculations should be made for stairs serving basement storeys and stairs serving upper storeys.

The population, P, should be divided by the number of available stairs.

The formula is useful to determine the width of stairs where people are not distributed evenly – either within a storey or between storeys.

In the formula, 200W represents the number of people estimated to have left the stair after 2.5 minutes of evacuation, and 50 (W – 0.3)(N – 1) represents the number of people estimated to be on the stair after 2.5 minutes of evacuation.

Worked examples

A 14 storey building contains 12 storeys of offices (ground + 11). The top two storeys contain flats that are served by separate stairs. What is the minimum width needed for the stairs that serve the office floors, for simultaneous evacuation? In the 11 above-ground-floor offices, 1200 people use the stairs. (People in the ground floor offices do not use the stairs.) In this example, two stairs are shown to satisfy the travel distance limitations.

a. The population is distributed evenly

The top office storey is at a height greater than 18m, therefore both stairs need lobby protection (see paragraph 3.34). Because both stairs are entered at each level via a protected lobby, both stairs can be assumed to be available (see paragraph 3.15).

$P = 1200/2 = 600, N = 11$

From the formula:

$600 = 200W + 50 (W - 0.3)(11 - 1)$

$600 = 200W + (50W - 15)(10)$

$600 = 200W + 500W - 150$

$750 = 700W$

$W = 1070mm$

Therefore both stairs should be at least 1070mm wide. But this needs to be increased to 1100mm, because the formula applies to stairs 1100mm wide or wider (see paragraph 3.18).

This width will also be adequate when one storey exit is discounted as described in paragraph 2.21. It also complies with paragraph 3.10a (i.e. the stair widths are not less than the minimum widths needed for 110 people in Table 2.3).

b. The population is not distributed evenly

(e.g. 1000 people occupy floors 1 to 9, and 200 occupy floors 10 and 11).

The top office storey is at a height greater than 18m, therefore both stairs need lobby protection (see paragraph 3.34). Because both stairs are entered at each level via a protected lobby, both stairs can be assumed to be available (see paragraph 3.15).

To find the width of:

- the stairs serving floors 10 and 11:

P = 200/2 = 100, N = 2

From the formula:

100 = 200W + 50 (W – 0.3)(2 – 1)

100 = 200W + (50W – 15)(1)

100 = 200W + 50W – 15

115 = 250W

W = 460mm

Therefore both stairs between the 9th floor landing and the top floor should be at least 460mm wide. But this needs to be increased to 1100mm, because the formula applies to stairs 1100mm wide or wider (see paragraph 3.18).

This width will also be adequate when one storey exit is discounted as described in paragraph 2.21. It also complies with paragraph 3.10a (i.e. the stair widths are not less than the minimum widths needed for 100 people in Table 2.3).

- the stairs serving floors 1 to 9:

P = 1200/2 = 600, N = 9

From the formula:

600 = 200W + 50 (W – 0.3)(9 – 1)

600 = 200W + (50W – 15)(8)

600 = 200W + 400W – 120

720 = 600W

W = 1200mm

Therefore both stairs between the ninth floor landing and the ground floor should be at least 1200mm wide.

This width will also be adequate when one storey exit is discounted as described in paragraph 2.21. It also complies with paragraph 3.10a (i.e. the stair widths are not less than the minimum widths needed for 134 people in Table 2.3).

Phased evacuation

3.19 Phased evacuation cannot be used in every type of building, but can be advantageous for escape stairs in high buildings. It requires supporting facilities, such as fire detection and alarm systems, to be provided and maintained.

In a phased evacuation, the first people to be evacuated are those with reduced mobility and those on the storey most immediately affected by the fire. If needed, subsequent evacuation is done two floors at a time, reducing disruption in large buildings.

Phased evacuation enables stairs to be narrower than with simultaneous evacuation, and may be used for any building provided it is not identified in paragraph 3.16.

3.20 Phased evacuation in buildings over 30m in height introduces the potential for escaping people to impede firefighters entering and operating within the building. This can be addressed by consulting with the fire and rescue service about special management procedures.

In very tall buildings, typically over 45m in height, physical measures may need to be incorporated, such as by discounting a stair.

3.21 A building (or part of a building) designed for phased evacuation should satisfy all of the following criteria.

a. At each storey except a top storey, stairs should be approached through a protected lobby or protected corridor.

b. At each storey, the lifts should be approached through a protected lobby (see paragraph 5.37).

c. Every floor should be a compartment floor (REI depending on height and use of the building).

d. If there is a storey with a floor over 30m above ground level, the building should be protected throughout by an automatic sprinkler system in accordance with Appendix E.

e. The building should be fitted with an appropriate fire warning system conforming to at least the L3 standard given in **BS 5839-1**.

f. An internal speech communication system should provide communication between a control point at fire and rescue service access level and a fire warden on every storey. The recommendations for phased evacuation provided in **BS 5839-1** should be followed. Where it is deemed appropriate to install a voice alarm, this should be in accordance with **BS 5839-8**.

3.22 The minimum width of stairs needed for phased evacuation is given in Table 3.3.

Table 3.3 Minimum width of stairs designed for phased evacuation

Maximum number of people in any storey	Stair width (mm)
100	1000
120	1100
130	1200
140	1300
150	1400
160	1500
170	1600
180	1700
190	1800

NOTES:

1. This table assumes a phased evacuation of the fire floor first followed by evacuation of not more than two floors at a time.
2. Unless a central handrail is provided, stairs with a rise of more than 30m should be a maximum width of 1400mm (see paragraph 3.10).
3. As an alternative to using this table, the minimum width (in mm) may be calculated from:
   $(P \times 10) - 100$
   where P = the number of people on the most heavily occupied storey.
   However, the minimum width of a stair should be 1000mm.

Worked example using Table 3.3

What is the minimum width needed for the stairs serving a 15 storey office building (ground + 14 office floors), assuming a total population of 2500 people (excluding the ground floor population, which does not use the stairs)? To satisfy the travel distance limitations, three stairs are required.

The building is over 45m in height and designed for phased evacuation. It has been decided to discount one stair to take account of fire and rescue service operations as described in paragraph 3.20. Therefore:

- Number of people per storey = 2500/14 = 179.

Each remaining stair must be able to accommodate half the population of one storey (i.e. 90 people).

Thus each stair should be 1000mm wide (maximum capacity 100 people).

This width will also be adequate when one storey exit is discounted as described in paragraph 2.21. It also complies with paragraph 3.10a (i.e. the stair widths are not less than the minimum width needed for 90 people in Table 2.3).

- At least one of those stairs needs to be a firefighting stair, therefore a minimum width of 1100mm is needed (see note 2 to Table 3.1).

Additional worked example using Table 3.3

What is the minimum width needed for the stairs serving a 9 storey office building (ground + 8 office floors), assuming a total population of 1920 people (excluding the ground floor population, which does not use the stairs)? To satisfy the travel distance limitations, two stairs are required.

As both stairs need to be entered at each level through a protected lobby (see paragraph 3.21), both stairs can be assumed to be available (see paragraph 3.15). Therefore:

- Number of people per storey = 1920/8 = 240.
- Each stair must be able to accommodate half the population of one storey (i.e. 240/2 = 120 people).
- Thus both stairs would require a width of 1100mm (maximum capacity 120 people) according to Table 3.3, but:
- Each storey exit needs to be able to serve 240 people, because of discounting as described in paragraph 2.21. The minimum exit width needed for 240 people in Table 2.3 is 1200mm. As described in paragraph 3.10a, the stair should be at least as wide as the storey exit serving it.
- The required stair width is therefore 1200mm.

Design and protection of escape stairs

Enclosure of escape stairs

3.23 Every internal escape stair should be a protected stairway (within a fire resisting enclosure). If it is also a protected shaft or firefighting shaft, additional guidance in Sections 8 and 17 applies.

There is one exception: an unprotected stair (e.g. an accommodation stair) may form part of an internal route to a storey exit or final exit, provided that the distance of travel and the number of people involved are very limited. For example, small premises (Section 4) and raised storage areas (see paragraphs 7.6 and 7.7).

d. Stairs more than 6m in height should be protected from adverse weather. Protection should prevent the build-up of snow or ice but does not require full enclosure.

e. Glazing in areas of fire resisting construction should be fixed shut and fire resisting, in terms of integrity but not insulation (minimum E 30).

3.33 Access to an external escape stair may be via a flat roof, provided the flat roof meets the requirements of paragraphs 2.31 and 2.32.

See para 3.32

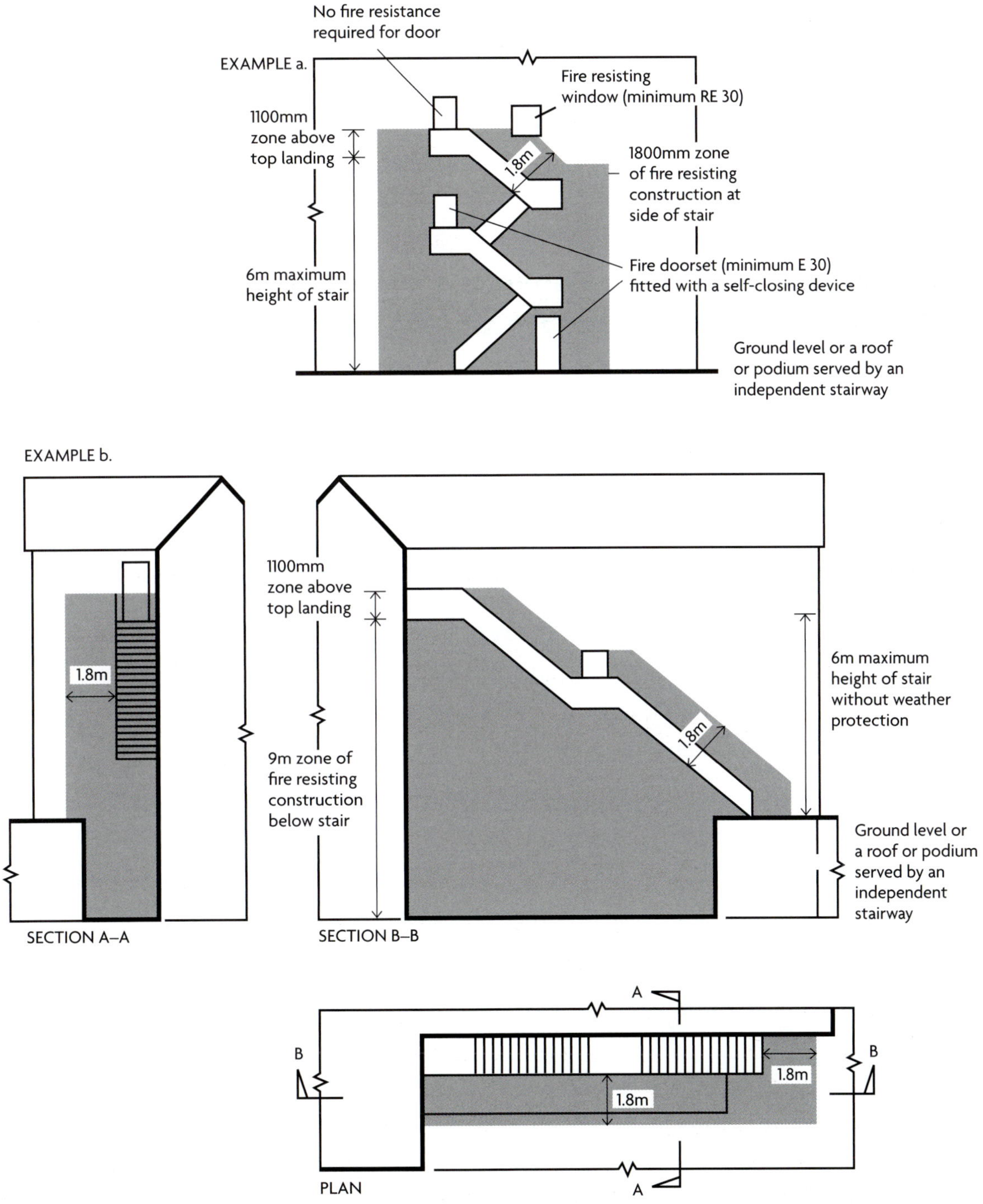

Diagram 3.4 Fire resistance of areas near to external stairs

Access lobbies and corridors

3.34 In the following situations, protected lobbies or protected corridors should be provided at all storeys above ground, except the top storey.

a. If the stair is the only one serving a building or part of a building that has more than one storey above or below the ground storey.

b. If the stair serves any storey at a height of 18m or more above ground level.

c. If the building is designed for phased evacuation.

d. If the stair is a firefighting stair.

e. If the option in paragraph 3.15b has been used so as not to discount one stair when calculating stair widths.

As an alternative to (a) to (c), a smoke control system as described in paragraph 3.15a may be used.

3.35 A protected lobby should be provided between an escape stair and a place of special fire hazard to protect from the ingress of smoke. The lobby should have a minimum $0.4m^2$ of permanent ventilation, or be protected by a mechanical smoke control system.

Exits from protected stairways

3.36 Every protected stairway should lead to a final exit, either directly or via an exit passageway. Any protected exit corridor or stair should have the same standard of fire resistance and lobby protection as the stair it serves. The exit from a protected stairway should comply with paragraphs 5.20 to 5.24.

Separation of adjoining protected stairways

3.37 The construction separating two adjacent protected stairways (or exit passageways) leading to different final exits should be imperforate.

Use of space within protected stairways

3.38 A protected stairway may only include any of the following.

a. Sanitary accommodation or washrooms, as long as the accommodation is not used as a cloakroom. A gas water heater or sanitary towel incinerator may be installed in the accommodation, but no other gas appliance.

b. *If the protected stairway is not a firefighting stair:* a lift well.

c. *If the protected stairway is not the only stair serving the building or part of the building: a* reception desk or enquiry office area at ground or access level. The reception or enquiry office area should have a maximum area of $10m^2$.

d. *If the protected stairway is not the only stair serving the building or part of the building:* cupboards enclosed with fire resisting construction.

Gas service and installation pipes in protected stairways

3.39 Gas service and installation pipes and meters should not be within a protected stairway, unless installed in accordance with the Pipelines Safety Regulations 1996 and the Gas Safety (Installation and Use) Regulations 1998.

Basement stairs

3.40 An escape stair forming part of the only escape route from an upper storey should *not* continue down to serve a basement storey. The basement storey should be served by a separate escape stair.

3.41 Where multiple escape stairs serve the upper storeys, only one needs to end at ground level. Other stairs may connect with the basement storeys if there is a protected lobby or a protected corridor between the stairs and accommodation at each basement level.

Section 4: Small premises

4.1 A 'small premises' is generally limited both in its size and in its number of occupants. When undivided, all of its parts are likely to be clearly visible to occupants. Occupants of small premises will be able to reach an exit quickly in an emergency and therefore a reduction in the number of exits and stairs is acceptable.

This guidance is not applicable to small premises where highly flammable materials are sold, stored or used.

4.2 Small premises should meet all of the following general conditions.

a. i. It should be single occupancy.

ii. It should not comprise more than a basement storey, ground storey and first storey.

iii. No storey should have a floor area more than $280m^2$.

b. Any kitchen or other open cooking arrangements should be at the extremity of any dead end remote from the exits.

c. For a bar or restaurant, the seating or standing accommodation (Table D1) should be planned for a maximum of 30 people per storey. The seating or standing accommodation for the ground storey may be planned for 100 people if it has a final exit independent of the stair.

4.3 The following paragraphs *only* apply in place of those provisions elsewhere in this Approved Document which relate to the following.

a. The number and position of exits and protected stairways.

b. Measuring distances of travel.

c. Open escape stairs.

For provisions other than those listed above, the guidance elsewhere in this Approved Document should be followed.

Construction

4.4 Except in kitchens, ancillary offices and stores, floor areas should be undivided so exits are clearly visible from all parts.

4.5 Store rooms should be enclosed with fire resisting construction (minimum REI 30).

Travel distance and number of escape routes

4.6 Escape routes should be sited so that the travel distance from any point of a storey to the nearest storey exit does not exceed the distance given in Table 4.1 (see Diagrams 4.1, 4.2 and 4.3). The siting of two or more exits or stairs should give effective alternative directions of travel from any point in a storey.

Table 4.1 Maximum travel distances in small premises

Storey	Maximum travel distance (m)
Ground storey with a single exit	27
Basement or first storey with a single stair	18
Storey with more than one exit/stair	45

NOTES:

If the internal layout of partitions, fittings, etc. is not known, direct distances, rather than travel distances, should be assessed. The direct distance should be assumed to be two-thirds of the travel distance.

The travel distance in small premises with an open stair is measured as follows.

a. In a basement: to the foot of the stair.

b. On a first storey: to the head of the stair.

See para 4.6

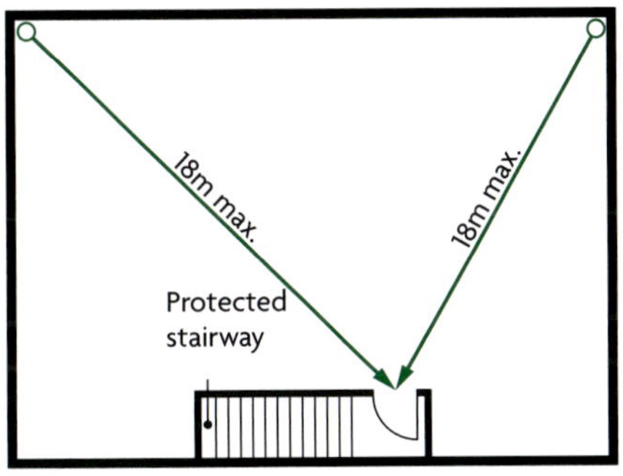

Basement and/or first storey

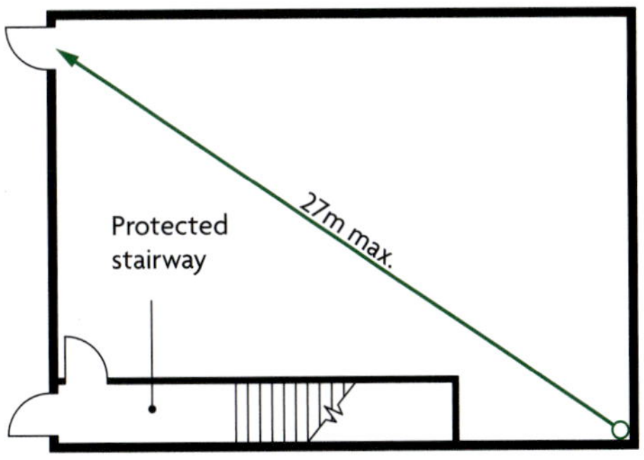

Ground storey

NOTE: Maximum floor area in any one storey 280m^2. Restricted accommodation if used as a restaurant or bar.

Diagram 4.1 Maximum travel distances in a small two or three storey premises with a single protected stairway to each storey

See para 4.6

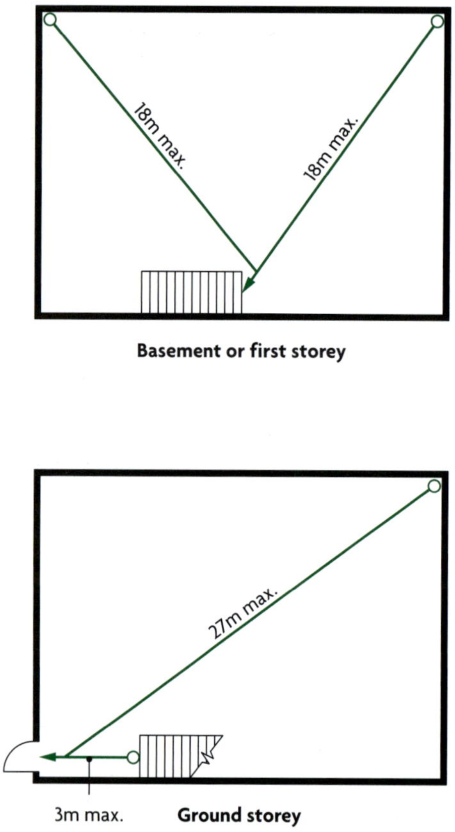

NOTES:

1. Maximum floor area in any one storey $90m^2$.
2. The premises may not be used as a resturant or bar.
3. Only acceptable in two storey premises (first + ground storey or basement + ground storey).
4. Travel distances are set out in Table 4.1.

Diagram 4.2 Maximum travel distances in a small three storey premises with a single open stair

See para 4.6

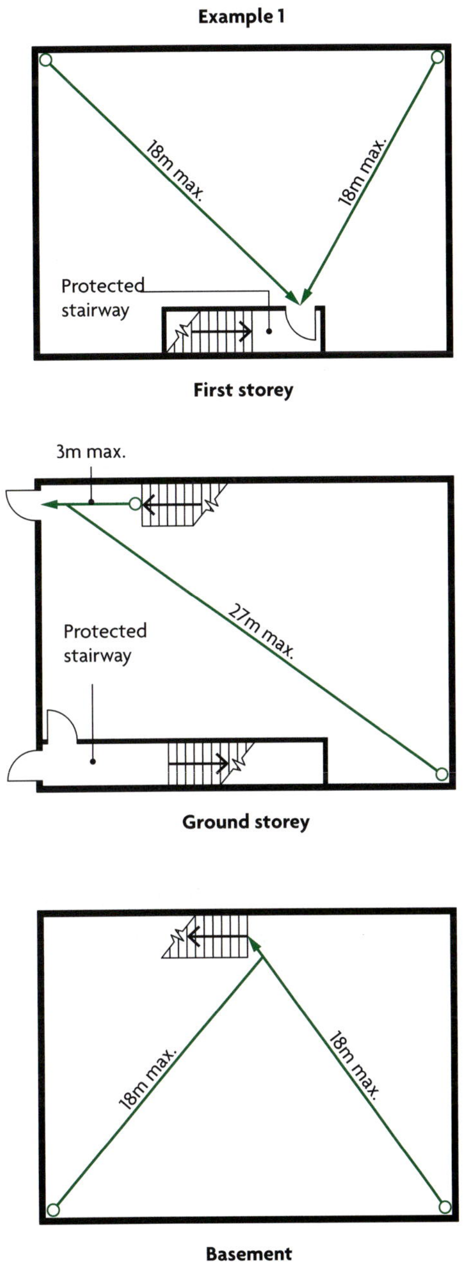

NOTES:

1. Maximum floor area in any one storey 90m^2.
2. Enclosed stair at ground storey level may be from either the basement or the first storey.
3. The premises may not be used as a resturant or bar.
4. Travel distances are set out in Table 4.1.

Diagram 4.3 Maximum travel distances in a small three storey premises with a single stair to each storey

Escape stairs in small premises

4.7 A single escape stair may be used in small premises.

4.8 An open stair may be used as a means of escape if all of the following apply.

a. The stair connects a maximum of two storeys.

b. The stair enters the ground storey a maximum of 3m from the final exit (see Diagrams 4.2 and 4.3).

c. The premises is not a bar or restaurant.

d. Either of the following applies.

i. The storey is also served by a protected stairway.

ii. The stair is a single stair and the floor area of any single storey is a maximum of $90m^2$.

4.9 Where the premises contains three storeys and a single open stair serves a top or bottom storey, the stair serving the other storey should be enclosed with fire resisting construction at the ground storey level and discharge to a final exit independent of the ground storey (see Diagram 4.3).

Section 5: General provisions

Introduction

5.1 This section applies to all buildings and deals with the design, construction and protection of escape routes and service installations.

Protection of escape routes

Fire resistance of enclosures

5.2 Fire resistance test criteria are set out in Appendix B. Standards of performance are summarised in Tables B3 and B4. Apart from specific situations described in Sections 1 and 2, and requirements B3 and B5, a minimum performance of REI 30 is sufficient to protect means of escape.

Fire resistance of doors

5.3 Fire resistance test criteria are set out in Appendix C. Standards of performance are summarised in Table C1.

Fire resistance of glazed elements

5.4 If glazed elements in fire resisting enclosures and doors can only meet the required integrity performance, their use is limited. These limitations depend on whether the enclosure forms part of a protected shaft (see Section 8) and the provisions set out in Appendix B, Table B5. If both integrity and insulation performance can be met, there is no restriction in this document on the use or amount of glass.

5.5 Glazed elements should also comply with the following, where necessary.

a. If the enclosure forms part of a protected shaft: Section 8.

b. Appendix B, Table B5.

c. Guidance on the safety of glazing: Approved Document K.

Doors on escape routes

5.6 Doors should be readily openable to avoid undue delay to people escaping. Doors on escape routes (both within and from the building) should comply with paragraphs 5.7 to 5.15. Guidance on door closing and 'hold open' devices for fire doorsets is set out in Appendix C.

Door fastenings

5.7 In general, doors on escape routes (whether or not the doors are fire doorsets) should be either of the following.

a. Not fitted with a lock, latch or bolt fastenings.

b. Fitted only with simple fastenings that are all of the following.

i. Easy to operate; it should be apparent how to undo the fastening.

ii. Operable from the side approached by people escaping.

iii. Operable without a key.

iv. Operable without requiring people to manipulate more than one mechanism.

Doors may be fitted with hardware to allow them to be locked when rooms are empty.

In places such as hotel bedrooms, locks may be fitted that are key operated from the outside and manually opened from the inside.

If a secure door is operated by code or combination keypad, swipe or proximity card, biometric data, etc., a security mechanism override should be possible from the side approached by people escaping.

5.8 Electrically powered locks should return to the unlocked position in all of the following situations.

a. If the fire detection and alarm system operates.

b. If there is loss of power or system error.

c. If the security mechanism override is activated.

Security mechanism overrides for electrically powered locks should be a Type A call point as described in **BS 7273-4**. The call point should be positioned on the side approached by people escaping. If the door provides escape in either direction, a call point should be installed on both sides of the door.

5.9 In places of assembly and shop and commercial buildings (purpose groups 4 and 5), doors on escape routes from rooms with more than 60 people should be either of the following.

a. *Not* fitted with locks, latches or bolts.

b. Fitted with panic fastenings in accordance with **BS EN 1125**.

In non-residential buildings (purpose groups 3 to 7), some final exit doors feature security locks that are used only when the building is empty. Such locks may be appropriate, but management procedures must emphasise their safe use.

5.10 Guidance on door closing and 'hold open' devices for fire doorsets is set out in Appendix C.

Direction of opening

5.11 The door of any doorway or exit should be hung to open in the direction of escape whenever reasonably practicable. It should always be hung to open in the direction of escape if either of the following conditions applies.

a. More than 60 people might be expected to use it during a fire.

b. There is a very high risk of fire with potential for rapid fire growth, such as with some industrial activities.

Amount of opening and effect on associated escape routes

5.12 All doors on escape routes should be hung to meet both of the following conditions.

a. Open by a minimum of 90 degrees.

b. Open with a swing that complies with both of the following.

i. Is clear of any change of floor level, other than a threshold or single step on the line of the doorway.

ii. Does not reduce the effective width of any escape route across a landing.

5.13 Any door opening towards a corridor or a stair should be recessed to prevent its swing encroaching on the effective width.

Vision panels in doors

5.14 Doors should contain vision panels in both of the following situations.

a. Where doors on escape routes divide corridors.

b. Where doors are hung to swing both ways.

Approved Document M contains guidance about vision panels in doors across accessible corridors and Approved Document K contains guidance about the safety of glazing.

Revolving and automatic doors

5.15 Where revolving doors, automatic doors and turnstiles are placed across escape routes they should comply with one of the following.

a. They are automatic doors of the required width and comply with one of the following conditions.

   i. Their failsafe system provides outward opening from any open position.

   ii. They have a monitored failsafe system to open the doors if the mains electricity supply fails.

   iii. They failsafe to the open position if the power fails.

b. Non-automatic swing doors of the required width are provided immediately adjacent to the revolving or automatic door or turnstile.

General provisions

Headroom in escape routes

5.16 Escape routes should have a minimum clear headroom of 2m. The only projections allowed below this height are door frames.

Flooring of escape routes

5.17 Escape route floor finishes should minimise their slipperiness when wet. Finishes include the treads of steps and surfaces of ramps and landings.

Ramps and sloping floors

5.18 A ramp forming part of an escape route should meet the provisions in Approved Document M. Any sloping floor or tier should have a pitch of not more than 35 degrees to the horizontal.

5.19 Guidance for where there is fixed seating is given in both of the following.

a. Approved Documents K and M give guidance on the design of ramps and associated landings, and on aisles and gangways where there is fixed seating.

b. Section 2 of this document refers to Annex D of **BS 9999**, which gives guidance on the design of means of escape in places with fixed seating.

Final exits

5.20 The width of a final exit should be at least the same as the minimum required width of the escape route it serves.

5.21 People should be able to rapidly leave the area around the building. Direct access to a street, passageway, walkway or open space should be available. The route away from the building should comply with the following.

a. Be well defined.

b. If necessary, have suitable guarding.

5.22 Final exits should not present a barrier for disabled people. Where the route to a final exit does not include stairs, a level threshold and, where necessary, a ramp should be provided.

5.23 Final exit locations should be clearly visible and recognisable.

5.24 Final exits should avoid outlets of basement smoke vents and openings to transformer chambers, refuse chambers, boiler rooms and similar risks.

Table 5.1 Provisions for escape lighting

Use of the building or part of the building	Areas requiring escape lighting
Residential	All common escape routes[1]
Office, industrial, storage and other non-residential	a. Underground or windowless accommodation b. Stairs either: • in a central core • that serve storey(s) more than 18m above ground level c. Internal corridors more than 30m long d. Open-plan areas of more than $60m^2$
Shop and commercial, and car parks	a. Underground or windowless accommodation b. Stairs either: • in a central core • that serve storey(s) more than 18m above ground level c. Internal corridors more than 30m long d. Open-plan areas of more than $60m^2$ e. All escape routes (other than the following exception) to which the public are admitted.[1] The exception is shops that meet all of the following: • have a maximum of three storeys • have no sales floor of more than $280m^2$ • are *not* a restaurant or bar
Assembly and recreation	a. All escape routes[1] b. Accommodation except for that which is open on one side to view sport or entertainment during normal daylight hours
Any purpose group	a. All toilet accommodation with a minimum floor area of $8m^2$ b. Electricity and generator rooms c. Switch room/battery room for emergency lighting system d. Emergency control rooms

NOTE:

1. Including external escape routes.

Lighting of escape routes

5.25 All escape routes should have adequate artificial lighting. If the mains electricity power supply fails, escape lighting should illuminate the routes listed in Table 5.1.

5.26 Escape stair lighting should be on a separate circuit from the electricity supply to any other part of the escape route.

5.27 Escape lighting should conform to **BS 5266-1**.

Exit signs

5.28 Every doorway or other exit providing access to a means of escape, other than exits in ordinary use (e.g. main entrances), should be distinctively and conspicuously marked by an exit sign in accordance with **BS ISO 3864-1** and **BS 5499-4**.

Advice on fire safety signs, including emergency escape signs, is given in the HSE publication *Safety Signs and Signals: Guidance on Regulations.*

Some buildings may require additional signs to comply with other legislation.

Protected power circuits

5.29 To limit potential damage to cables in protected circuits, all of the following should apply.

a. Cables should be sufficiently robust.

b. Cable routes should be carefully selected and/or physically protected in areas where cables may be exposed to damage.

c. Methods of cable support should be class A1 rated and offer at least the same integrity as the cable. They should maintain circuit integrity and hold cables in place when exposed to fire.

5.30 A protected circuit to operate equipment during a fire should achieve all of the following.

a. Cables should achieve PH 30 classification when tested in accordance with **BS EN 50200** (incorporating Annex E) or an equivalent standard.

b. It should only pass through parts of the building in which the fire risk is negligible.

c. It should be separate from any circuit provided for another purpose.

5.31 Guidance on cables for large and complex buildings is given in **BS 5839-1**, **BS 5266-1** and **BS 8519**.

Lifts

Evacuation lifts

5.32 Generally, lifts should not be used when there is a fire in the building, unless their use forms part of a management plan for evacuating people and the following conditions are met.

a. Lifts are appropriately sited and protected.

b. Lifts contain safety features to ensure they remain usable during a fire.

Guidance on the design and use of evacuation lifts is given in Annex G to **BS 9999**.

5.33 Where a firefighting lift is provided, it can be used to evacuate disabled people. Management plans should describe how this would be managed, and what will happen when the fire and rescue service arrives.

Fire protection of lift installations

5.34 Lift wells should comply with one of the following conditions.

a. Be sited within the enclosures of a protected stairway.

b. Be enclosed with fire resisting construction (minimum REI 30) when in a position that might prejudice the means of escape.

5.35 A lift well connecting different compartments should form a protected shaft (see Section 8).

5.36 Lifts that rise within a large volume such as a mall or atrium and do not have a conventional well, such as wall-climber or feature lifts, may be at risk if they run through a smoke reservoir. Care should be taken to maintain the integrity of the smoke reservoir and protect people in the lift.

5.37 In buildings designed for phased evacuation or progressive horizontal evacuation, if the lift well is not within the enclosures of a protected stairway, its entrance should be separated at every storey by a protected lobby (minimum REI 30).

5.38 In basements and enclosed car parks, the lift should be within the enclosure of a protected stairway. Otherwise, the lift should be approached only via a protected lobby or protected corridor (minimum REI 30).

5.39 If a lift delivers into a protected corridor or protected lobby serving sleeping accommodation and also serves a storey containing a high fire risk (such as a kitchen, communal areas, stores, etc.) then the lift should be separated from the high fire risk area(s) by a protected lobby or protected corridor (minimum REI 30).

5.40 A lift shaft serving storeys above ground level should not serve any basement if either of the following applies.

a. There is only one escape stair serving storeys above ground level and smoke from a basement fire would adversely affect escape routes in the upper storeys.

b. The lift shaft is within the enclosure to an escape stair that terminates at ground level.

5.41 Lift machine rooms should be sited over the lift well where possible. Where buildings or part of a building with only one stairway make this arrangement impractical, the lift machine room should be sited outside the protected stairway.

Refuse chutes and storage

5.42 Refuse storage chambers, refuse chutes and refuse hoppers should be sited and constructed in accordance with **BS 5906**.

5.43 Refuse chutes and rooms for storing refuse should meet both of the following conditions.

a. Be separated from other parts of the building by fire resisting construction (minimum REI 30 in buildings with a top storey up to 5m above ground level; otherwise REI 60).

b. Not be situated within a protected stairway or protected lobby.

5.44 The approach to rooms containing refuse chutes or for storing refuse should comply with one of the following conditions.

a. Be directly from the open air.

b. Be through a protected lobby with a minimum $0.2m^2$ of permanent ventilation.

5.45 Access openings to refuse storage chambers should *not* be sited next to escape routes or final exits.

Shop store rooms

5.46 Fully enclosed walk-in store rooms should be separated from retail areas with fire resisting construction (minimum REI 30) if they negatively affect the means of escape. The fire resisting construction is not necessary if the walk-in store room complies with either of the following.

a. Has an automatic fire detection and alarm system.

b. Is fitted with sprinklers.

B2

Requirement B2: Internal fire spread (linings)

This section deals with the following requirement from Part B of Schedule 1 to the Building Regulations 2010.

Requirement

Requirement	*Limits on application*
Internal fire spread (linings) **B2.** (1) To inhibit the spread of fire within the building, the internal linings shall— (a) adequately resist the spread of flame over their surfaces; and (b) have, if ignited, either a rate of heat release or a rate of fire growth, which is reasonable in the circumstances. (2) In this paragraph "internal linings" means the materials or products used in lining any partition, wall, ceiling or other internal structure.	

Intention

In the Secretary of State's view, requirement B2 is met by achieving a restricted spread of flame over internal linings. The building fabric should make a limited contribution to fire growth, including a low rate of heat release.

It is particularly important in circulation spaces, where linings may offer the main means by which fire spreads and where rapid spread is most likely to prevent occupants from escaping.

Requirement B2 *does not* include guidance on the following.

a. Generation of smoke and fumes.

b. The upper surfaces of floors and stairs.

c. Furniture and fittings.

Section 6: Wall and ceiling linings

Classification of linings

6.1 The surface linings of walls and ceilings should meet the classifications in Table 6.1.

Table 6.1 Classification of linings

Location	Classification
Small rooms of maximum internal floor area: a. $4m^2$ in residential accommodation b. $30m^2$ in non-residential accommodation	D-s3, d2
Other rooms (including garages)	C-s3, d2
Other circulation spaces	B-s3, d2[1]

NOTE:

1. Wallcoverings which conform to **BS EN 15102**, achieving at least class C-s3, d2 and bonded to a class A2-s3, d2 substrate, will also be acceptable.

Walls

6.2 For the purposes of this requirement, a wall includes both of the following.

a. The internal surface of internal and external glazing (except glazing in doors).

b. Any part of a ceiling which slopes at an angle greater than 70 degrees to the horizontal.

6.3 For the purposes of this requirement, a wall *does not* include any of the following.

a. Doors and door frames.

b. Window frames and frames in which glazing is fitted.

c. Architraves, cover moulds, picture rails, skirtings and similar narrow members.

d. Fireplace surrounds, mantle shelves and fitted furniture.

6.4 Parts of walls in rooms may be of lower performance than stated in Table 6.1, but no worse than class D-s3, d2. In any one room, the total area of lower performance wall lining should be less than an area equivalent to half of the room's floor area, up to a maximum of:

a. $20m^2$ in residential accommodation.

b. $60m^2$ in non-residential accommodation.

Ceilings

6.5 For the purposes of this requirement, a ceiling includes all of the following.

a. Glazed surfaces.

b. Any part of a wall at 70 degrees or less to the horizontal.

c. The underside of a gallery.

d. The underside of a roof exposed to the room below.

6.6 For the purposes of this requirement, a ceiling *does not* include any of the following.

a. Trap doors and their frames.

b. The frames of windows or rooflights and frames in which glazing is fitted.

c. Architraves, cover moulds, picture rails, exposed beams and similar narrow members.

Fire resisting ceilings

6.7 The need for cavity barriers in concealed floor or roof spaces can be reduced by installing a fire resisting ceiling (minimum EI 30) below the cavity, complying with Diagram 9.3.

Rooflights

6.8 Rooflights should meet the following classifications, according to material. No guidance for European fire test performance is currently available, because there is no generally accepted test and classification procedure.

a. Non-plastic rooflights should meet the relevant classification in Table 6.1.

b. Plastic rooflights, if the limitations in Table 6.2 and Table 14.2 are observed, should be a minimum class D-s3, d2 rating. Otherwise they should meet the relevant classification in Table 6.1.

Special applications

6.9 Any flexible membrane covering a structure, other than an air-supported structure, should comply with Appendix A of **BS 7157**.

6.10 Guidance on the use of PTFE-based materials for tension-membrane roofs and structures is given in BRE report BR 274.

Fire behaviour of insulating core panels used internally

6.11 Insulating core panels consist of an inner core of insulation sandwiched between, and bonded to, a membrane, such as galvanised steel or aluminium.

Where they are used internally they can present particular problems with regard to fire spread, and should meet all of the following conditions.

a. Panels should be sealed to prevent exposure of the core to a fire. This includes at joints and where services penetrate the panel.

b. In high fire risk areas, such as kitchens, places of special fire hazard, or in proximity to where hot works occur, only class A1 cored panels should be used.

c. Fixing systems for all panels should be designed to take account of the potential for the panel to delaminate. For instance, where panels are used to form a suspended ceiling, the fixing should pass through the panel and support it from the lower face.

Other controls on internal surface properties

6.12 Guidance on the control of flame spread is given in the following sections.

a. Stairs and landings: Section 3 (escape stairs) and Section 17 (firefighting shafts).

b. Section 9: exposed surfaces above fire-protecting suspended ceilings.

Thermoplastic materials

General provisions

6.13 Thermoplastic materials that do not meet the classifications in Table 6.1 can be used as described in paragraphs 6.14 to 6.18. No guidance for European fire test performance is currently available, because there is no generally accepted test and classification procedure.

Thermoplastic materials are defined in Appendix B, paragraph B11. Classifications used here are explained in paragraph B13.

Windows

6.14 Thermoplastic material classified as a TP(a) rigid product may be used to glaze external windows to rooms, *but not* external windows to circulation spaces. Approved Document K includes guidance on the safety of glazing.

Rooflights

6.15 In rooms and circulation spaces other than protected stairways, rooflights may be constructed of thermoplastic material if they comply with both of the following.

a. The lower surface is classified as TP(a) rigid or TP(b).

b. The size and location of the rooflights follow the limits in Table 6.2 and in Table 14.2 and Table 14.3.

Lighting diffusers

6.16 The following paragraphs apply to lighting diffusers forming part of a ceiling. Diffusers may be part of a luminaire or used below sources of light. The following paragraphs *do not* apply to diffusers of light fittings attached to the soffit of a ceiling or suspended beneath a ceiling (Diagram 6.1).

See para 6.16

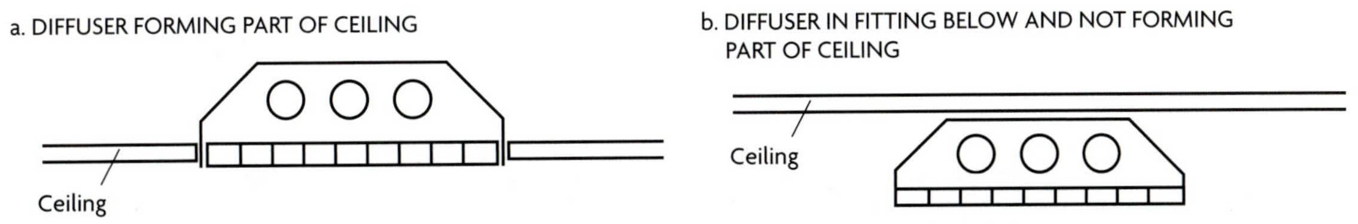

Diagram 6.1 Lighting diffuser in relation to ceiling

6.17 Diffusers constructed of thermoplastic material may be incorporated in ceilings to rooms and circulation spaces, but not to protected stairways, if both of the following conditions are met.

a. Except for the upper surfaces of the thermoplastic panels, wall and ceiling surfaces exposed in the space above the suspended ceiling should comply with paragraph 6.1.

b. Diffusers should be classified as one of the following.

i. TP(a) rigid – no restrictions on their extent.

ii. TP(b) – limited in their extent (see Table 6.2 and Diagram 6.2).

Suspended or stretched-skin ceilings

6.18 A ceiling constructed from TP(a) flexible panels should meet the following conditions.

a. Have a maximum area of $5m^2$.

b. Be supported on all sides.

See para 6.17

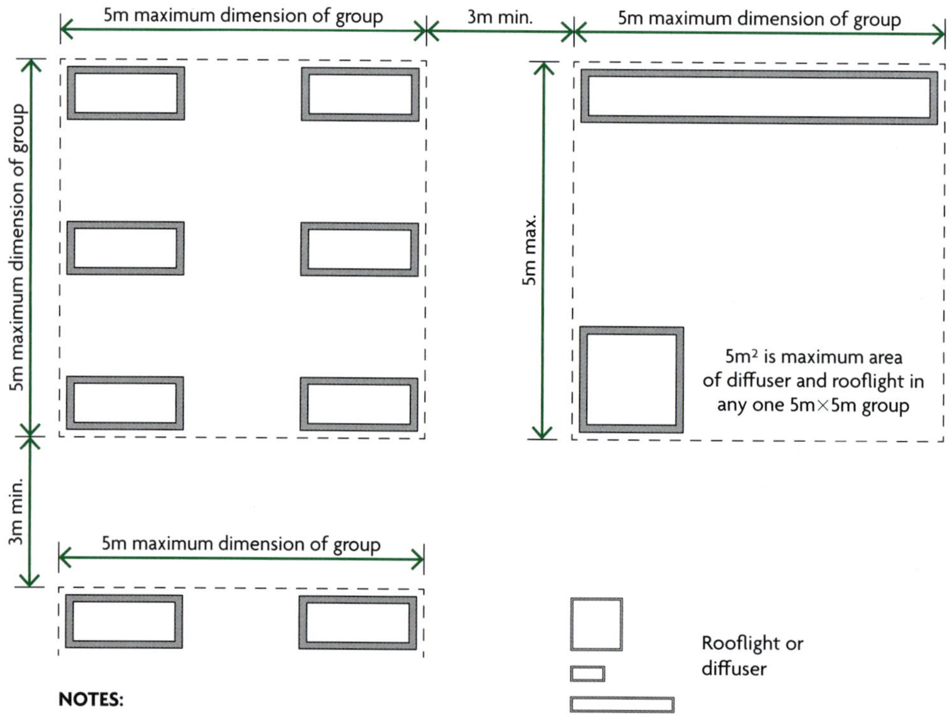

NOTES:

1. Upper and lower surfaces of suspended ceiling, between plastic panels, to comply with paragraph 6.1.

2. No restriction on class D-s3, d2 diffusers or rooflights in small rooms.

3. See note 4 to Table 6.2.

Diagram 6.2 Layout restrictions on class D-s3, d2 plastic rooflights, TP(b) rooflights and TP(b) lighting diffusers

Table 6.2 Limitations applied to thermoplastic rooflights and lighting diffusers in suspended ceilings and class D-s3, d2 plastic rooflights[1]

Minimum classification of lower surface	Use of space below the diffusers or rooflight	Maximum area of each diffuser or rooflight[2] (m^2)	Maximum total area of diffusers and rooflights as a percentage of floor area of the space in which the ceiling is located (%)	Minimum separation distance between diffusers or rooflights[2] (m)
TP(a)	Any except protected stairway	No limit[3]	No limit	No limit
D-s3, d2[4] or TP(b)	Rooms	1	50[5][6]	A distance equal to the largest plan dimension of the largest diffuser or rooflight (see Diagram 6.3)
		5	50[5][6]	3[6]
	Circulation spaces except protected stairways	5	15[5]	3

NOTES:

1. This table does not apply to products that meet the provisions in Table 6.1.
2. Smaller rooflights and diffusers can be grouped together provided that both of the following satisfy the dimensions in Diagram 6.2 or 6.3.
   a. The overall size of the group.
   b. The space between one group and any others.
3. Lighting diffusers of TP(a) flexible rating should be used only in panels of a maximum of $5m^2$ each. See paragraph 6.18.
4. There are no limits on the use of class D-s3, d2 materials in small rooms. See Table 6.1.
5. The minimum 3m separation given in Diagram 6.2 between each $5m^2$ group must be maintained. Therefore, in some cases, it may not be possible to use the maximum percentage quoted.
6. Class D-s3, d2 rooflights to rooms in industrial and other non-residential purpose group buildings (purpose groups 3 to 7) may be spaced 1800mm apart provided both of the following conditions are met.
   a. The rooflights are evenly distributed.
   b. The total area of the rooflights does not exceed 20% of the area of the room.

See Table 6.2

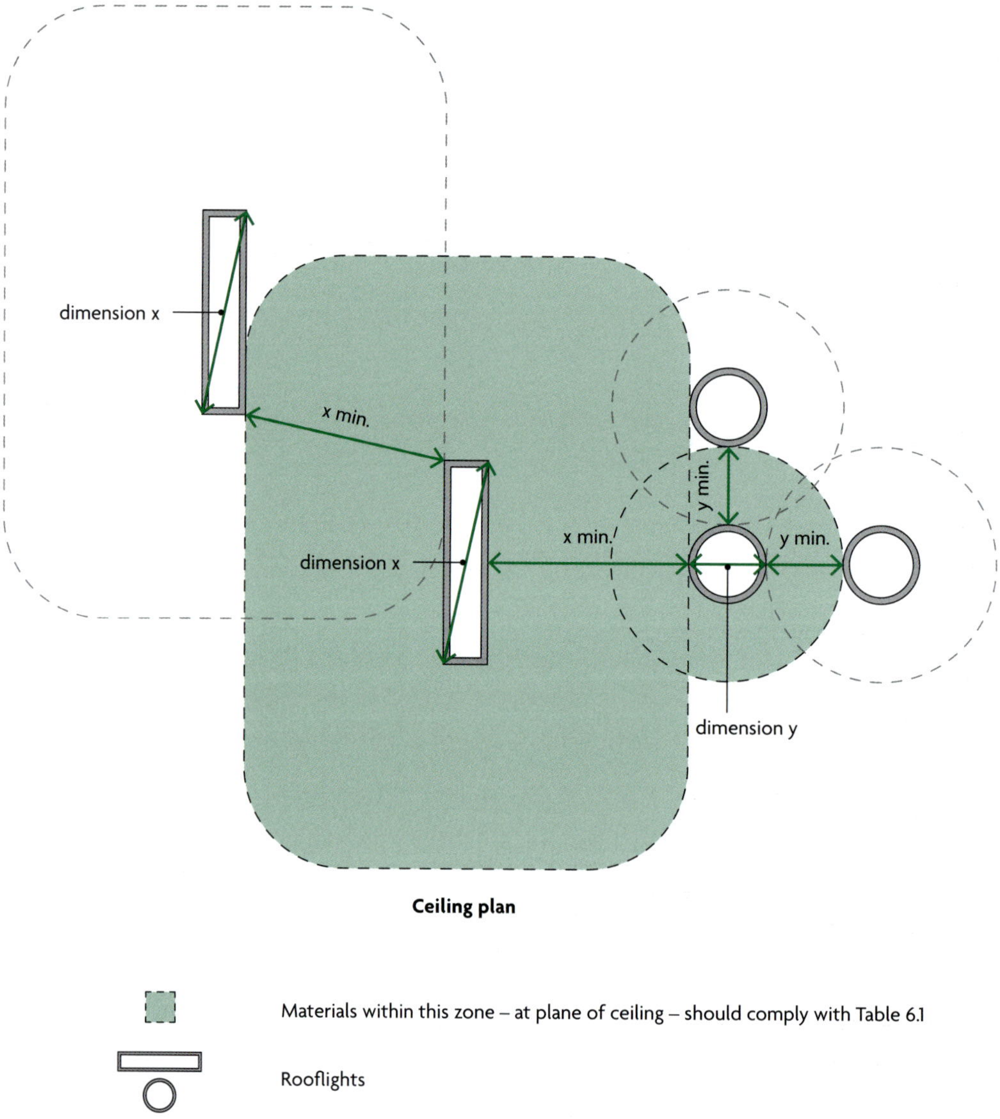

Diagram 6.3 Layout restrictions on small class D-s3, d2 plastic rooflights, TP(b) rooflights and TP(b) lighting diffusers

Requirement B3: Internal fire spread (structure)

These sections deal with the following requirement from Part B of Schedule 1 to the Building Regulations 2010.

Requirement

Requirement	*Limits on application*
Internal fire spread (structure)	
B3. (1) The building shall be designed and constructed so that, in the event of fire, its stability will be maintained for a reasonable period	
(2) A wall common to two or more buildings shall be designed and constructed so that it adequately resists the spread of fire between those buildings. For the purposes of this sub-paragraph a house in a terrace and a semi-detached house are each to be treated as a separate building.	
(3) Where reasonably necessary to inhibit the spread of fire within the building, measures shall be taken, to an extent appropriate to the size and intended use of the building, comprising either or both of the following— (a) sub-division of the building with fire-resisting construction; (b) installation of suitable automatic fire suppression systems.	Requirement B3(3) does not apply to material alterations to any prison provided under section 33 of the Prison Act 1952.
(4) The building shall be designed and constructed so that the unseen spread of fire and smoke within concealed spaces in its structure and fabric is inhibited.	

Intention

In the Secretary of State's view, requirement B3 is met by achieving all of the following.

a. For defined periods, loadbearing elements of structure withstand the effects of fire without loss of stability.

b. Compartmentation of buildings by fire resisting construction elements.

c. Automatic fire suppression is provided where it is necessary.

d. Protection of openings in fire-separating elements to maintain continuity of the fire separation.

e. Inhibition of the unseen spread of fire and smoke in cavities, to reduce the risk of structural failure and spread of fire and smoke, where they pose a threat to the safety of people in and around the building.

The extent to which any of these measures are necessary is dependent on the use of the building and, in some cases, its size, and on the location of the elements of construction.

Section 7: Loadbearing elements of structures

Fire resistance standard

7.1 Elements such as structural frames, beams, columns, loadbearing walls (internal and external), floor structures and gallery structures should have, as a minimum, the fire resistance given in Appendix B, Table B3.

7.2 Appendix B includes guidance on all of the following.

a. Provisions to ensure that where one element of structure supports or stabilises another element of structure, the supporting element has no less fire resistance than the other element (see Table B4).

b. Measures so that elements common to more than one building or compartment are constructed to the standard of the more onerous of the relevant provisions.

c. Special provisions about fire resistance of elements of structure in single storey buildings.

d. Concessions in respect of fire resistance of elements of structure in basements where one or more sides of the basement are open at ground level.

Exclusions from the provisions for elements of structure

7.3 The following are excluded from the definition of 'element of structure'.

a. A structure that supports only a roof, unless either of the following applies.

   i. The roof performs the function of a floor, such as for parking vehicles, or as a means of escape.

   ii. The structure is essential for the stability of an external wall that needs to be fire resisting (e.g. to achieve compartmentation or for the purposes of preventing fire spread between buildings).

b. The lowest floor of the building.

c. A platform floor.

d. A loading gallery, fly gallery, stage grid, lighting bridge or any gallery provided for similar purposes or for maintenance and repair.

e. External walls, such as curtain walls or other forms of cladding, which transmit only self weight and wind loads and do not transmit floor load.

NOTE: In some cases, structural members within a roof may be essential for the structural stability system of the building. In these cases, the structural members in the roof do not just support a roof and must demonstrate the relevant fire resistance for the building as required by paragraph 7.2a above.

Additional guidance

7.4 If a loadbearing wall is any of the following, guidance in other sections may also apply.

a. A compartment wall (including a wall common to two buildings): Section 8.

b. Enclosing a place of special fire hazard: Section 8, paragraph 8.7.

c. Protecting a means of escape: Sections 2 to 5.

d. An external wall: Sections 12 and 13.

e. Enclosing a firefighting shaft: Section 17.

7.5 If a floor is also a compartment floor, see Section 8.

Raised storage areas

7.6 The normal provisions for fire resistance may be too onerous to apply to raised, free-standing floors (sometimes supported by racking) in single storey buildings used for industrial and storage purposes. The introduction of raised storage areas can alter the effective number of storeys in the building (see the definition of 'storey' in Appendix A).

7.7 A structure that does not have the minimum fire resistance specified in Appendix B, Table B4, is acceptable if it satisfies all of the following conditions.

a. The structure meets both of the following conditions.

  i. It has only one tier.

  ii. It is used for storage purposes *only*.

b. The people likely to be on the floor at any one time are both of the following.

  i. Few in number.

  ii. Not members of the public.

c. The floor is open above and below to the room or space in which it is situated.

d. The means of escape from the floor is in accordance with Sections 2 to 5.

e. The floor meets both of the following conditions.

  i. It is not more than 10m in width or length.

  ii. It is a maximum of half the floor area of the space in which it is situated.

  The limitations in (e) may be adjusted if any of the following apply.

f. If the lower level has an automatic fire detection and alarm system meeting the recommendations of **BS 5839-1**, then the floor size may be increased to not more than 20m in either width or length.

g. If agreed with the building control body and the fire and rescue service, then it may be possible to vary this dimension and area. However, the safety of firefighters and the distance they may need to travel over or under the floor must be considered.

h. If the building is fitted throughout with an automatic sprinkler system complying with Appendix E, then no limits are set for the size of the floor.

Section 8: Compartmentation/sprinklers

Provision of compartmentation

All purpose groups

8.1 All of the following should be provided as compartment walls and compartment floors and should have, as a minimum, the fire resistance given in Appendix B, Table B3.

8.2 A wall common to two or more buildings should be a compartment wall.

8.3 Parts of a building occupied mainly for different purposes should be separated from one another by compartment walls and/or compartment floors. Compartmentation is not needed if one of the different purposes is ancillary to the other. See paragraphs 0.23 and 0.24.

8.4 Effective compartmentation relies on both of the following.

a. Fire resistance should be continuous at the join between elements forming a compartment.

b. Any openings between two compartments should not reduce the fire resistance.

8.5 The lowest floor in a building does not need to be a compartment floor.

Protected shafts

8.6 Stairs and service shafts connecting compartments should be protected to restrict the spread of fire between the compartments. These are called protected shafts. Walls or floors surrounding a protected shaft are considered to be compartment walls or compartment floors.

Places of special fire hazard

8.7 Fire resisting construction enclosing these places should achieve minimum REI 30. These walls and floors are not compartment walls and compartment floors.

'Residential (institutional)' buildings including health care

8.8 All floors should be constructed as compartment floors.

8.9 Paragraphs 2.35 and 2.36 give guidance on the provisions for compartment walls in care homes that use progressive horizontal evacuation.

'Residential (other)' buildings

8.10 In 'residential (other)' (purpose group (2(b)) buildings, all floors should be compartment floors.

Non-residential buildings

8.11 In buildings in a non-residential purpose group (purpose groups 3 to 7), the following should be compartment walls and compartment floors.

a. Every wall needed to divide the building to observe the compartment size limits in Table 8.1 (Diagram 8.1a).

b. Every floor, *if* the building or separated part of the building (see paragraph 8.19) has a top storey that is more than 30m above ground level (Diagram 8.1b).

c. The floor of the ground storey, *if* the building has one or more basements (Diagram 8.1c), except in small premises (see paragraph 4.2).

d. The floor of every basement storey (except the lowest floor), *if* the building or separated part has a basement more than 10m below ground level (Diagram 8.1d).

e. If the building comprises 'shop and commercial', 'industrial' or 'storage' premises (purpose groups 4, 6, 7): every wall or floor dividing a building into separate occupancies (spaces used by different organisations, whether they fall within the same purpose group or not).

f. See also the provision in paragraph 5.46 for store rooms in shops to be separated from retail areas by fire resisting construction (minimum REI 30).

8.12 In two storey 'shop and commercial' or 'industrial' buildings (purpose groups 4 or 6), where the use of the upper storey is ancillary to the use of the ground storey, the ground storey may be treated as a single storey building for fire compartmentation purposes where all of the following apply.

a. The area of the upper storey does not exceed the lower of:

   i. 20% of the area of the ground storey

   ii. $500m^2$.

b. The upper storey is compartmented from the lower one.

c. The upper storey has a means of escape independent of the lower storey escape routes.

Every place of special fire hazard (see Appendix E) should be enclosed with fire resisting construction.

Buildings containing one or more atria

8.13 Detailed advice on atria in buildings is given in Annexes B and C of **BS 9999**. For the purposes of this document, **BS 9999** applies *only where the atrium breaches a compartment*.

See para 8.11

A. EXAMPLE OF COMPARTMENTATION IN AN UNSPRINKLERED SHOP see paragraph 8.11a

None of the floors in this case would need to be compartment floors, but the two storeys exceeding 2000m² would need to be divided into compartments a maximum of 2000m² by compartment walls.

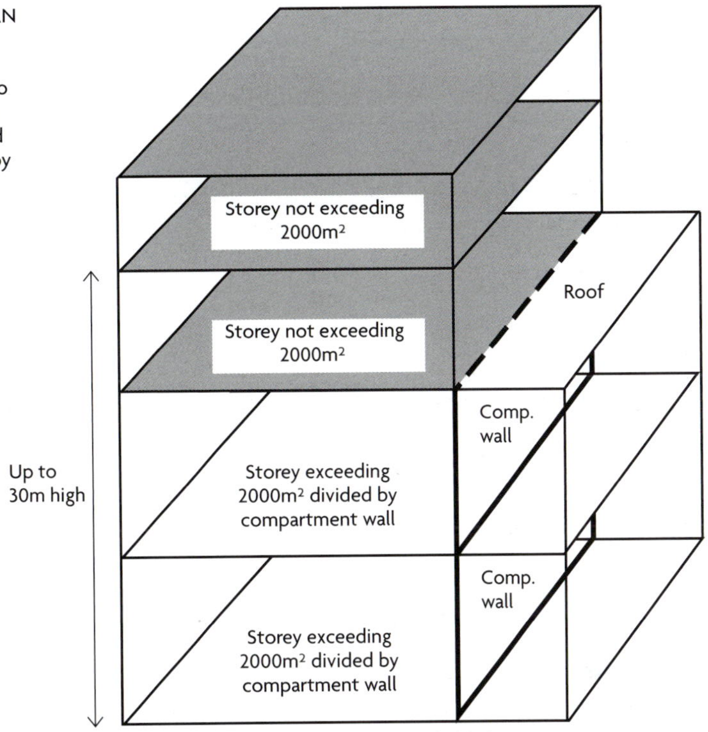

The compartment walls in example (a) do not need to be in one vertical plane

B. COMPARTMENTATION IN TALL BUILDINGS see paragraph 8.11b

In a building over 30m in height all storeys should be separated by compartment floors. For advice on the special conditions in atrium buildings see Annex B of BS 9999

C. SHALLOW BASEMENT see paragraph 8.11c

Only the floor of the ground storey need be a compartment floor if the lower basement is at a depth of not more than 10m

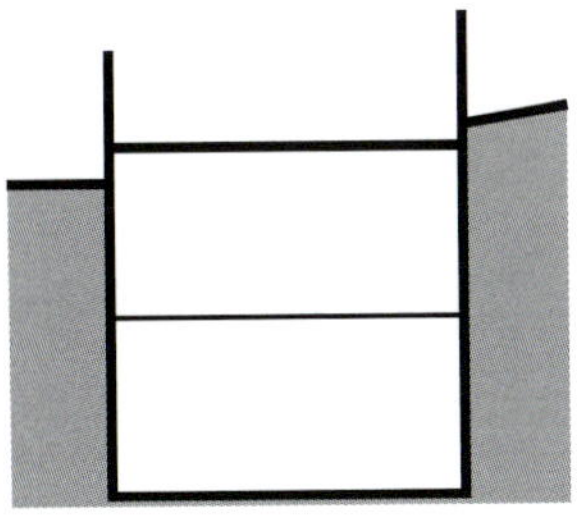

D. DEEP BASEMENTS see paragraph 8.11d

All basement storeys to be separated by compartment floors if any storey is at a depth of more than 10m

Diagram 8.1 Compartment floors: illustration of guidance in paragraph 8.11

Table 8.1 Maximum dimensions of building or compartment (non-residential buildings)

Purpose group of building or part	Height of floor of top storey above ground level (m)	Maximum floor area of any one storey in the building or any one storey in a compartment (m^2)		
		Single storey buildings		Multi-storey buildings
Office	No limit[1]	No limit		No limit
Assembly and recreation, shop and commercial:				
a. Shops – without sprinkler system	No limit[1]	2000		2000
Shops – with sprinkler system[2]	No limit	No limit		4000
b. Elsewhere – without sprinkler system	No limit[1]	No limit		2000
Elsewhere – with sprinkler system[2]	No limit	No limit		4000
Industrial[3] Without sprinkler system[1]	Not more than 18 More than 18	No limit N/A		7000 2000[4]
With sprinkler system[2]	Not more than 18 More than 18	No limit N/A		14,000 4000[4]
	Height of floor of top storey above ground level (m)	Maximum floor area (m^2)	Maximum height (m)[5]	Maximum compartment volume (m^3)
		Single storey buildings		Multi-storey buildings
Storage[3] and other non-residential:				
a. Car park for light vehicles	No limit	No limit	No limit	No limit
b. Any other building or part:				
Without sprinkler system[1]	Not more than 18 More than 18	20,000 N/A	18 N/A	20,000 4000[4]
With sprinkler system[2]	Not more than 18 More than 18	No limit	No limit	40,000 8000[4]

NOTES:

This table recommends that where the maximum size limitations placed on a building without a sprinkler system are exceeded, a sprinkler system in accordance with Appendix E should be provided.

1. See Appendix B, Table B4 for sprinkler system height requirements.
2. 'With sprinkler system' means that the building is fitted throughout with an automatic sprinkler system in accordance with Appendix E.
3. In certain industrial and storage uses that are subject to other legislation, for example the storage of LPG and certain chemicals, additional limitations on floor area and/or additional sprinkler provisions might apply.
4. This reduced limit applies only to storeys that are a minimum of 18m above ground level. Below this height the higher limit applies.
5. Compartment height is measured from finished floor level to the underside of the roof or ceiling.

Sprinklers

8.14 Buildings within the 'office', 'shop and commercial', 'assembly and recreation', 'industrial' and 'storage and other non-residential' (except car parks for light vehicles) purpose groups (purpose groups 3 to 7(a)) require sprinklers where there is a top storey above 30m. The sprinkler system should be provided in accordance with Appendix E.

Construction of compartment walls and compartment floors

General provisions

8.15 All compartment walls and compartment floors should achieve both of the following.

a. Form a complete barrier to fire between the compartments they separate.

b. Have the appropriate fire resistance, as given in Appendix B, Tables B3 and B4.

8.16 Timber beams, joists, purlins and rafters may be built into or carried through a masonry or concrete compartment wall if the openings for them are both of the following.

a. As small as practicable.

b. Fire-stopped.

If trussed rafters bridge the wall, failure of the truss due to a fire in one compartment should not cause failure of the truss in another compartment.

8.17 Where services could provide a source of ignition, the risk of fire developing and spreading into adjacent compartments should be controlled.

Compartment walls between buildings

8.18 Adjoining buildings should only be separated by walls, not floors. Compartment walls common to two or more buildings should comply with both of the following.

a. Run the full height of the building in a continuous vertical plane.

b. Be continued through any roof space to the underside of the roof.

Separated parts of buildings

8.19 Compartment walls forming a separated part of a building should run the full height of the building in a continuous vertical plane.

Separated parts can be assessed independently to determine the appropriate standard of fire resistance in each. The two separated parts can have different standards of fire resistance.

Other compartment walls

8.20 Compartment walls not described in paragraphs 8.18 and 8.19 should run the full height of the storey in which they are situated.

8.21 Compartment walls in a top storey beneath a roof should be continued through the roof space.

Junction of compartment wall or compartment floor with other walls

8.22 At the junction with another compartment wall or an external wall, the fire resistance of the compartmentation should be maintained. Fire-stopping that meets the provisions in paragraphs 10.24 to 10.29 should be provided.

8.23 At the junction of a compartment floor and an external wall with no fire resistance, the external wall should be restrained at floor level. The restraint should reduce movement of the wall away from the floor if exposed to fire.

8.24 Compartment walls should be able to accommodate deflection of the floor, when exposed to fire, by either of the following means.

a. Between the wall and floor, provide a head detail that is capable of maintaining its integrity while deforming.

b. Design the wall so it maintains its integrity by resisting the additional vertical load from the floor above.

Where compartment walls are located within the middle half of a floor between vertical supports, the deflection may be assumed to be 40mm unless a smaller value can be justified by assessment. Outside this area, the limit can be reduced linearly to zero at the supports.

For steel beams that do not have the required fire resistance, reference should be made to SCI Publication P288.

Junction of compartment wall with roof

8.25 A compartment wall should achieve both of the following.

a. Meet the underside of the roof covering or deck, with fire-stopping to maintain the continuity of fire resistance.

b. Be continued across any eaves.

8.26 To reduce the risk of fire spreading over the roof from one compartment to another, a 1500mm wide zone of the roof, either side of the wall, should have a covering classified as B_{ROOF}(t4), on a substrate or deck of a material rated class A2-s3, d2 or better, as set out in Diagram 8.2a.

Thermoplastic rooflights that, because of paragraph 14.7, are regarded as having a B_{ROOF}(t4) classification are *not* suitable for use in that zone.

8.27 Materials achieving class B-s3, d2 or worse used as a substrate to the roof covering and any timber tiling battens, fully bedded in mortar or other suitable material for the width of the wall (Diagram 8.2b), may extend over the compartment wall in buildings that are both of the following.

a. A maximum of 15m high.

b. In one of the following purpose groups.

i. All residential purpose groups (purpose groups 1 and 2) other than 'residential (institutional)' (purpose group 2(a)).

ii. 'Office' (purpose group 3).

iii. 'Assembly and recreation' (purpose group 5).

8.28 Double-skinned insulated roof sheeting should incorporate a band of material rated class A2-s3, d2 or better, a minimum of 300mm in width, centred over the wall.

8.29 As an alternative to the provisions of paragraph 8.26 or 8.27, the compartment wall may extend through the roof for a minimum of either of the following (see Diagram 8.2c).

a. Where the height difference between the two roofs is less than 375mm, 375mm above the top surface of the adjoining roof covering.

b. 200mm above the top surface of the adjoining roof covering where either of the following applies.

i. The height difference between the two roofs is 375mm or more.

ii. The roof coverings either side of the wall are of a material classified as B_{ROOF}(t4).

See paras 8.26 to 8.29

a. ANY BUILDING OR COMPARTMENT

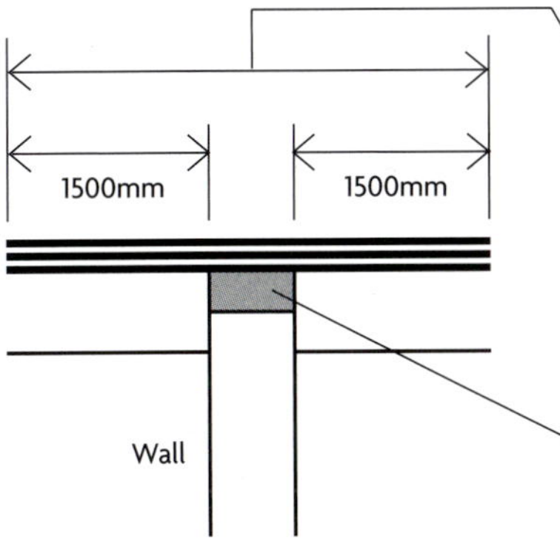

Roof covering over this distance to be designated B_{ROOF}(t4) rated on deck of material of class A2-s3, d2 or better. Roof covering and deck could be composite structure, e.g. profiled steel cladding.

Double-skinned insulated roof sheeting should incorporate a band of material rated class A2-s3, d2 or better, a minimum of 300mm in width, centred over the wall.

If roof support members pass through the wall, fire protection to these members for a distance of 1500mm on either side of the wall may be needed to delay distortion at the junction (see paragraph 8.16).

Fire-stopping to be carried up to underside of roof covering, e.g. roof tiles.

b. RESIDENTIAL (OTHER), OFFICE, OR ASSEMBLY AND RECREATION USE, AND NOT MORE THAN 15M HIGH

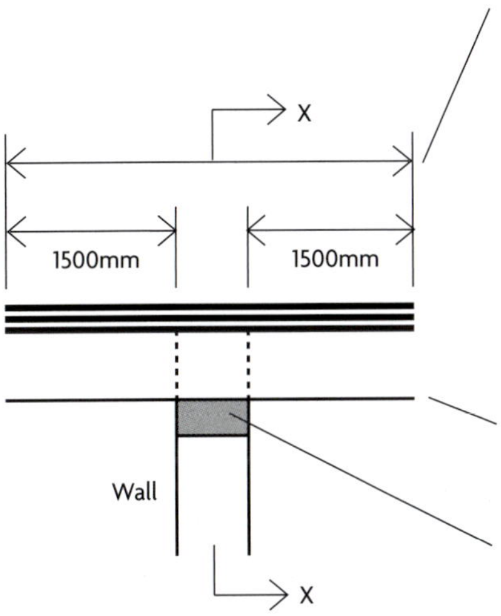

Roof covering to be designated B_{ROOF}(t4) rated for at least this distance.

Boarding (used as a substrate) or timber tiling battens may be carried over the wall provided that they are fully bedded in mortar (or other no less suitable material) where over the wall.

Thermoplastic insulation materials should not be carried over the wall.

Double-skinned insulated roof sheeting with a thermoplastic core should incorporate a band of material of class A2-s3, d2 at least 300mm wide centred over the wall.

Sarking felt may also be carried over the wall.

If roof support members pass through the wall, fire protection to these members for a distance of 1500mm on either side of the wall may be needed to delay distortion at the junction (see paragraph 8.16).

Fire-stopping to be carried up to underside of roof covering, boarding or slab.

Section X–X

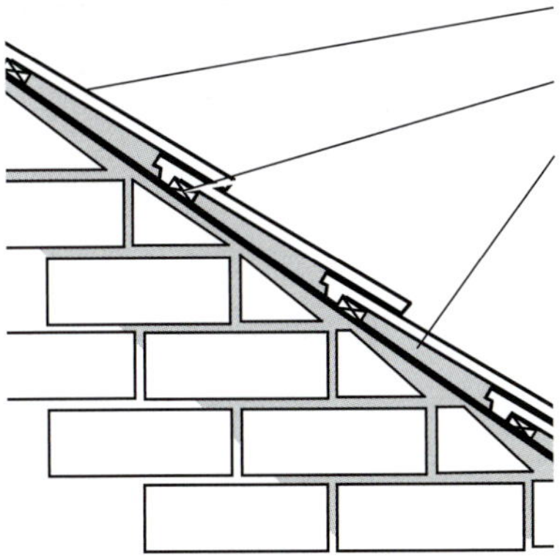

Roof covering to be designated B_{ROOF}(t4) rated for at least 1500mm either side of wall.

Roofing battens and sarking felt may be carried over the wall.

Fire-stopping to be carried up to underside of roof covering above and below sarking felt.

NOTES:

1. Fire-stopping should be carried over the full thickness of the wall.
2. Fire-stopping should be extended into any eaves.
3. The compartment wall does not necessarily need to be constructed of masonry.

c. ANY BUILDING OR COMPARTMENT

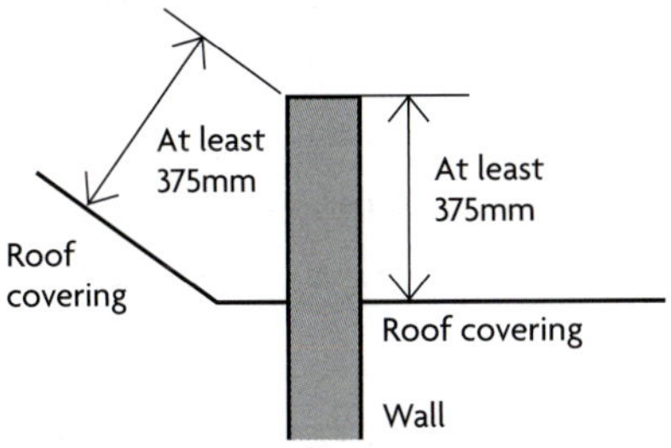

The wall should be extended up through the roof for a height of at least 375mm above the top surface of the adjoining roof covering.

Where there is a height difference of at least 375 mm between two roofs or where the roof coverings on either side of the wall are B_{ROOF}(t4) rated, the height of the upstand/parapet wall above the highest roof may be reduced to 200mm.

Diagram 8.2 Junction of compartment wall with roof

Openings in compartmentation

Openings in compartment walls separating buildings or occupancies

8.30 Openings in a compartment wall common to two or more buildings, or between different occupancies in the same building, should be limited to those for either of the following.

a. A fire doorset providing a means of escape, which has the same fire resistance as the wall and is fitted in accordance with the provisions in Appendix C.

b. The passage of a pipe that complies with the provisions in Section 10.

Openings in other compartment walls or in compartment floors

8.31 Openings should be limited to those for any of the following.

a. Fire doorsets fitted in accordance with the provisions in Appendix C.

b. Pipes, ventilation ducts, service cables, chimneys, appliance ventilation ducts or ducts encasing one or more flue pipes, complying with the provisions in Section 10.

c. Refuse chutes of class A1 construction.

d. Atria designed in accordance with Annexes B and C of **BS 9999**.

e. Protected shafts that conform to the provisions in the following paragraphs.

Protected shafts

8.32 Any stair or other shaft passing directly from one compartment to another should be enclosed in a protected shaft. Protected shafts should be used for the following only, but may also include sanitary accommodation and washrooms.

a. Stairs.

b. Lifts.

c. Escalators.

d. Chutes.

e. Ducts.

f. Pipes.

g. Additional provisions apply for both of the following.

i. Protected shafts that are protected stairways: Sections 2 to 5.

ii. Stairs that are also firefighting stairs: Section 17.

Construction of protected shafts

8.33 The construction enclosing a protected shaft (Diagram 8.3) should do all of the following.

a. Form a complete barrier to fire between the compartments connected by the shaft.

b. Have the appropriate fire resistance given in Appendix B, Table B3, *except for uninsulated glazed screens that meet the provisions of paragraph 8.34*.

c. Satisfy the provisions for ventilation and the treatment of openings in paragraphs 8.38 and 8.39.

See para 8.33

The diagram shows three common examples which illustrate the principles. The elements enclosing the shaft (unless formed by adjacent external walls) are compartment walls and floors.

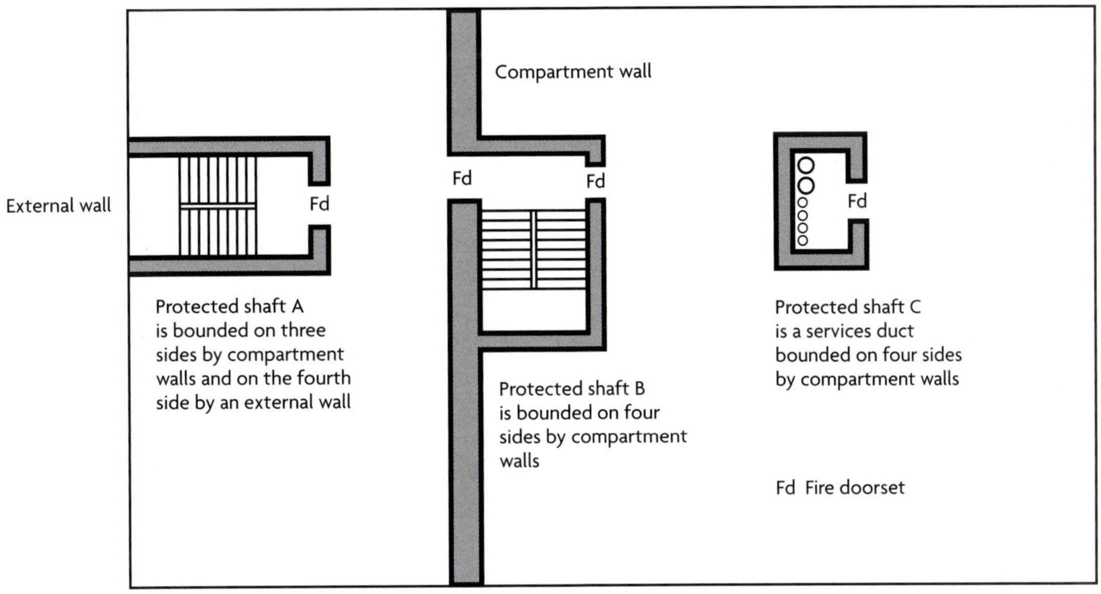

The shaft structure (including any openings) should meet the relevant provisions for: compartment walls (see paragraphs 8.15 to 8.31), external walls (see sections 12 and 13 and Diagram 3.3).

Diagram 8.3 Protected shafts

Uninsulated glazed screens to protected shafts

8.34 An uninsulated glazed screen may be incorporated in the enclosure to a protected shaft between a stair and a lobby or corridor entered from the stair. The enclosure must conform to Diagram 8.4 and meet all of the following conditions.

a. The standard of fire resistance required for the protected stairway is not more than REI 60.

b. The glazed screen complies with the following.

  i. It achieves a minimum rating of E 30.

  ii. It complies with the guidance on limits on areas of uninsulated glazing in Appendix B, Table B5.

c. The lobby or corridor is enclosed with fire resisting construction achieving a minimum rating of REI 30.

8.35 Where the measures in Diagram 8.4 are not provided, then both of the following apply.

a. The enclosing walls should comply with Appendix B, Table B3.

b. The doors should comply with Appendix B, Table B5.

See para 8.34

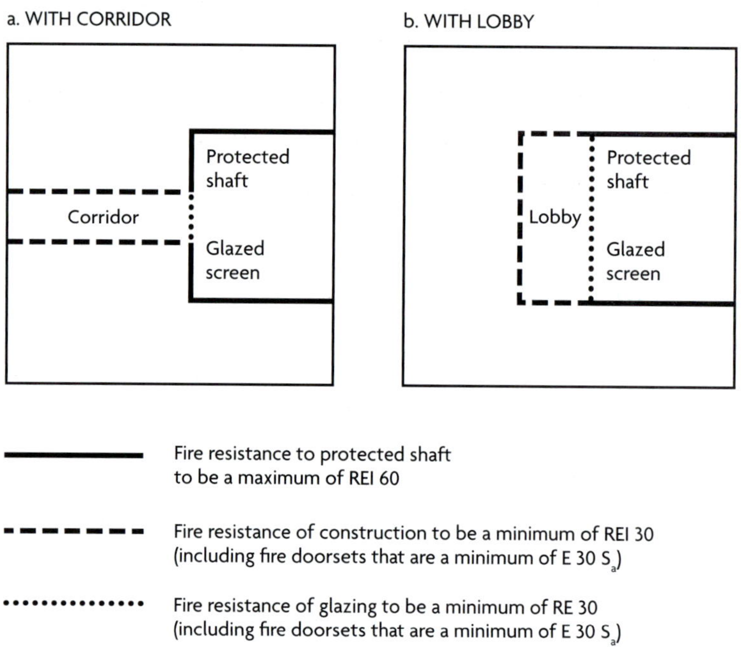

Diagram 8.4 Uninsulated glazed screen separating protected shaft from lobby or corridor

Pipes for oil or gas and ventilation ducts in protected shafts

8.36 A protected shaft containing a protected stairway and/or a lift *should not* also contain either of the following.

a. A pipe that conveys oil, other than in the mechanism of a hydraulic lift.

b. A ventilating duct. Two exceptions are as follows.

   i. A duct provided for pressurising the protected stairway to keep it smoke free.

   ii. A duct provided only to ventilate the protected stairway.

A pipe that is completely separated from a protected shaft by fire resisting construction is not considered to be contained within that shaft.

8.37 In a protected shaft, any pipe carrying natural gas or LPG should be both of the following.

a. Of screwed steel or all-welded steel construction.

b. Installed in accordance with both of the following.

   i. The Pipelines Safety Regulations 1996.

   ii. The Gas Safety (Installation and Use) Regulations 1998.

Ventilation of protected shafts conveying gas

8.38 A protected shaft conveying piped flammable gas should be ventilated direct to the outside air, by ventilation openings at high and low level in the shaft.

Any extension of the storey floor into the protected shaft should not compromise the free movement of air throughout the entire length of the shaft.

Guidance on shafts conveying piped flammable gas, including the size of ventilation openings, is given in **BS 8313**.

Openings into protected shafts

8.39 The external wall of a protected shaft does not normally need to have fire resistance. Situations where there are provisions are given in paragraph 3.29 (external walls of protected stairways, which may also be protected shafts) and paragraph 17.2 (firefighting shafts).

Openings in other parts of the enclosure to a protected shaft should be limited to the following.

a. If a wall common to two or more buildings forms part of the enclosure, only the following openings should be made in that wall.
   i. A fire doorset providing a means of escape, which has the same fire resistance as the wall and is fitted in accordance with the provisions in Appendix C.
   ii. The passage of a pipe that meets the provisions in Section 10.
b. Other parts of the enclosure (other than an external wall) should only have openings for any of the following.
   i. Fire doorsets of the appropriate fire resistance, fitted in accordance with the provisions in Appendix C.
   ii. The passage of pipes which meet the provisions in Section 10.
   iii. Inlets to, outlets from and openings for a ventilation duct (if the shaft contains or serves as a ventilating duct), meeting the provisions in Section 10.
   iv. The passage of lift cables into a lift machine room (if the shaft contains a lift). If the machine room is at the bottom of the shaft, the openings should be as small as practicable.

Section 9: Cavities

9.1 Cavities in the construction of a building provide a ready route for the spread of smoke and flame, which can present a greater danger as any spread is concealed. For the purpose of this document, a cavity is considered to be any concealed space.

See para 9.2

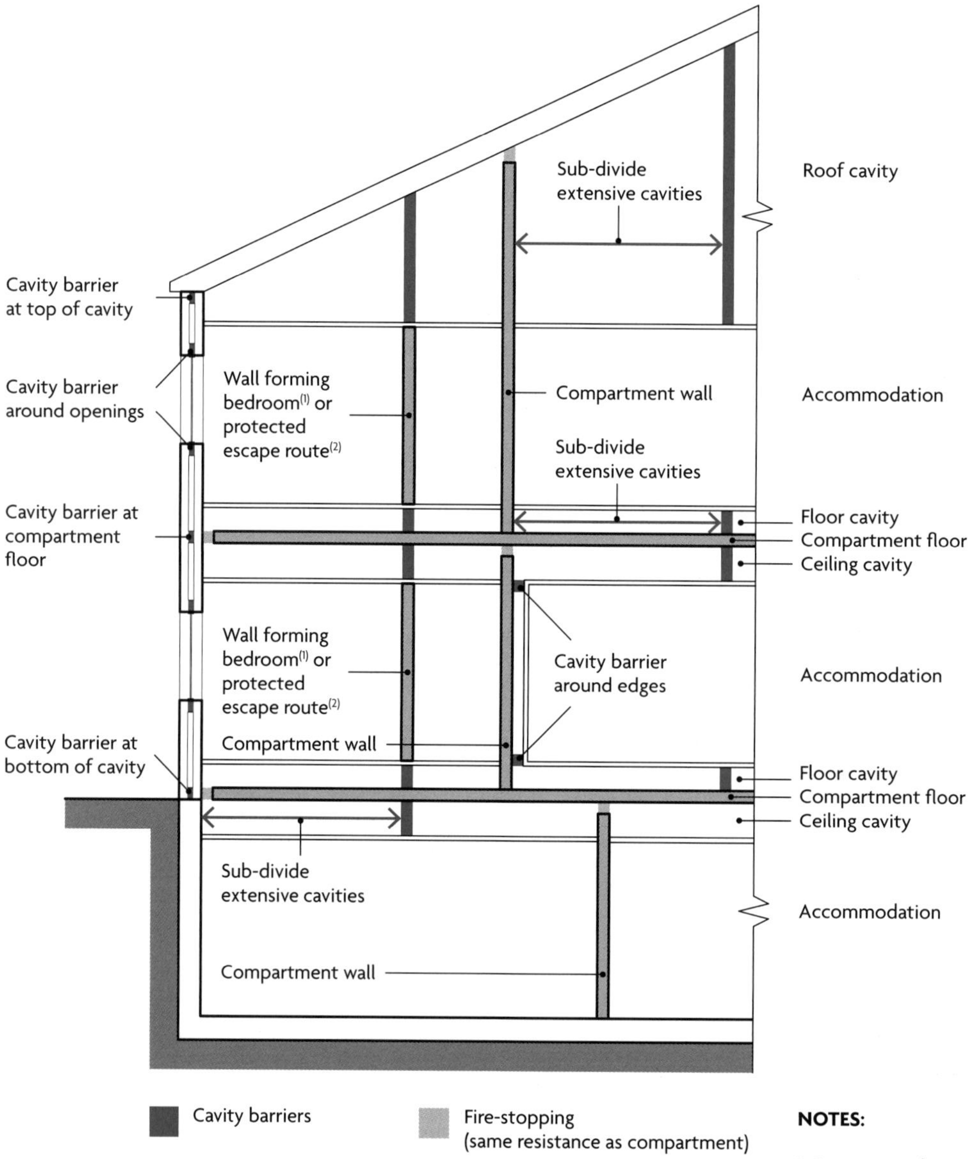

NOTES:

1. See paragraph 9.7.

2. See paragraph 9.5.

Diagram 9.1 Provisions for cavity barriers

Provision of cavity barriers

9.2 To reduce the potential for fire spread, cavity barriers should be provided for both of the following.

a. To divide cavities.

b. To close the edges of cavities.

See Diagram 9.1. Cavity barriers should not be confused with fire-stopping details (Section 10).

Pathways around fire-separating elements

Junctions and cavity closures

9.3 Cavity barriers should be provided at all of the following locations.

a. At the edges of cavities, including around openings (such as windows, doors and exit/entry points for services).

b. At the junction between an external cavity wall and every compartment floor and compartment wall.

c. At the junction between an internal cavity wall and every compartment floor, compartment wall or other wall or door assembly forming a fire resisting barrier.

This does not apply where a wall meets the conditions of Diagram 9.2.

9.4 It is not appropriate to complete a line of compartment walls by fitting cavity barriers above them. The compartment wall should extend to the underside of the floor or roof above.

Protected escape routes

9.5 If the fire resisting construction of a protected escape route is either of the following.

a. Not carried to full storey height.

b. At the top storey, not carried to the underside of the roof covering.

Then the cavity *above or below* the fire resisting construction should be either of the following.

i. Fitted with cavity barriers on the line of the enclosure.

ii. For cavities above the fire resisting construction, enclosed on the lower side by a fire resisting ceiling (minimum EI 30) that extends throughout the building, compartment or separated part (see Diagram 9.3).

Cavities affecting alternative escape routes

9.6 In divided corridors, cavity barriers may be needed to prevent alternative escape routes being affected by fire and/or smoke (see paragraph 2.27 and Diagram 2.9).

Separation of bedrooms in 'residential (institutional)' and 'residential (other)' buildings

9.7 Where a cavity exists above or below a partition between bedrooms because the enclosure is not carried to full storey height or to the underside of the roof covering, the guidance in paragraph 9.5 should be followed.

Double-skinned corrugated or profiled roof sheeting

9.8 Cavity barriers are not required between double-skinned corrugated or profiled insulated roof sheeting if the sheeting complies with all of the following.

a. The sheeting is rated class A2-s3, d2 or better.

b. Both surfaces of the insulating layer are rated class C-s3, d2 or better.

c. Both surfaces of the insulating layer make contact with the inner and outer skins of cladding (Diagram 9.4).

See para 9.3

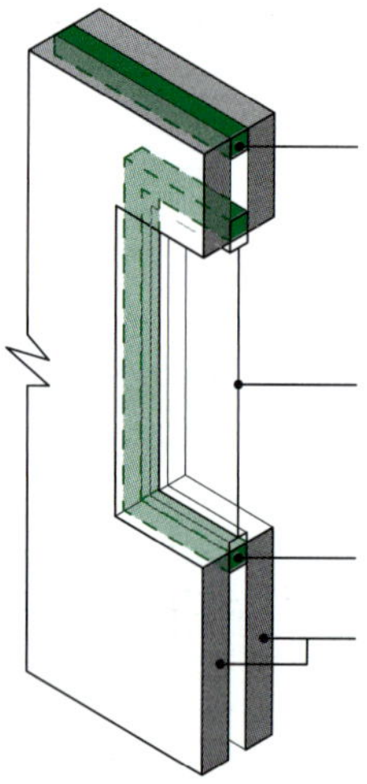

NOTES:

1. Materials used to close the cavity in this arrangement do not need to achieve a specific performance in relation to fire resistance.
2. Domestic meter cupboards may be installed provided that the following conditions are met:
   a. There are no more than two cupboards per dwelling
   b. The openings in the outer wall leaf are not bigger than 800×500mm for each cupboard
   c. The inner leaf is not penetrated except by a sleeve not more than 80×80mm, which is fire-stopped.
3. Materials achieving class B-s3, d2 or worse may be placed within the cavity.

Diagram 9.2 Cavity walls excluded from provisions for cavity barriers

See para 9.5

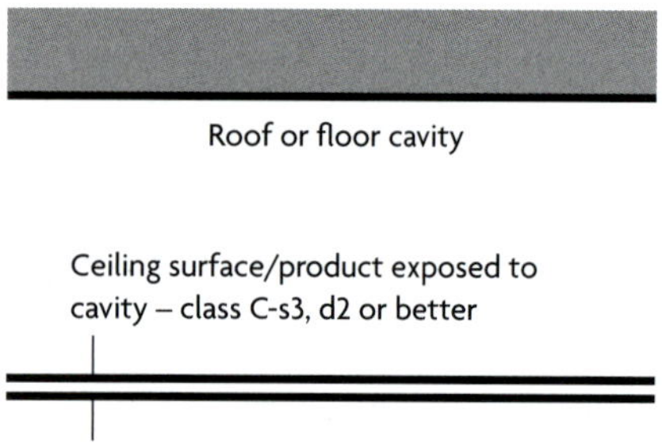

NOTE:

The ceiling should meet the following conditions.

a. Provide a minimum fire resistance of EI 30.
b. Be imperforate, except for an opening described in paragraph 9.17.
c. Extend throughout the building or compartment.
d. Not be easily demountable.

Diagram 9.3 Fire resisting ceiling below cavity

See para 9.8

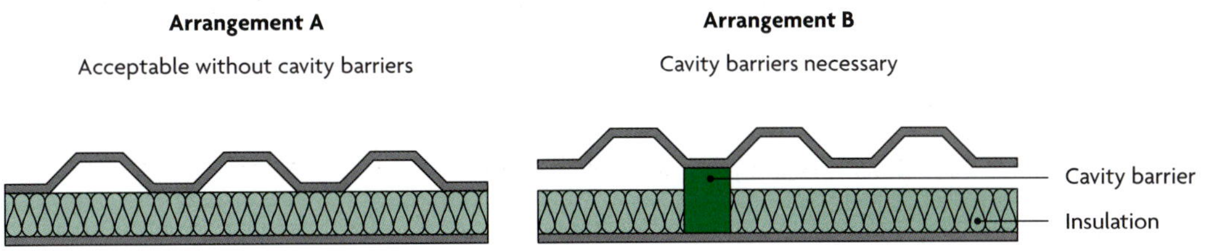

The insulation should make contact with both skins of sheeting. See also Diagram 8.2a regarding the need for fire-stopping where such roofs pass over the top of a compartment wall.

Diagram 9.4 Provisions for cavity barriers in double-skinned insulated roof sheeting

Extensive cavities

Maximum dimensions of cavities

9.9 Cavity barriers should be used to divide any cavity (including roof spaces). Table 9.1 sets out maximum dimensions for undivided cavities.

Table 9.1 Maximum dimensions of cavities in buildings other than dwellings (purpose groups 2 to 7)

Location of cavity	Class of surface/product exposed in cavity (excluding the surface of any pipe, cable or conduit, or any insulation to any pipe)	Maximum dimension in any direction (m)
Between roof and a ceiling	Any	20
Any other cavity	Class C-s3, d2 or better	20
	Worse than Class C-s3, d2	10

9.10 Table 9.1 does not apply to any of the following cavities.

a. A cavity in a wall that is fire resisting only because it is loadbearing.

b. A cavity in a wall that meets the conditions of Diagram 9.2.

c. A floor or roof cavity above a fire resisting ceiling (Diagram 9.3) that extends throughout the building or compartment to a maximum of 30m.

d. In a building not put to residential or institutional use, a cavity that does not contain materials achieving class B-s3, d2 or worse and is formed either:

   i. behind the external skin of an external cladding system with a masonry or concrete inner leaf a minimum of 75mm thick

   ii. by overcladding an existing masonry (or concrete) external wall or an existing concrete roof.

e. A cavity that meets the conditions of paragraph 9.8.

f. A cavity below a floor next to the ground or next to oversite concrete, if either:

   i. the cavity is less than 1000mm in height

   ii. the cavity is not normally accessible by people, unless there are openings in the floor such that it is possible for materials to accumulate in the cavity (in which case cavity barriers should be provided and access should be provided to the cavity for cleaning).

9.11 If a single room with a ceiling cavity or underfloor cavity exceeds the dimensions in Table 9.1, cavity barriers need only be provided on the line of the enclosing walls/partitions of that room, if both of the following apply.

a. The cavity barriers are a maximum of 40m apart.

b. The surface of the material/product exposed in the cavity is class C-s3, d2 or better.

9.12 If the undivided area exceeds 40m in one or both directions, there is no limit to its size if all of the following conditions are met.

a. Together, the room and cavity form a compartment separated from the rest of the building.

b. Both of the following apply.

   i. The building is fitted with an automatic fire detection and alarm system that conforms to **BS 5839-1**.

   ii. Detectors are only required in the cavity to satisfy **BS 5839-1**.

c. If the cavity is used as a plenum then the recommendations for recirculating air distribution systems in Section 32 of **BS 9999** are followed.

d. Both of the following apply.

   i. The exposed surface of the material/product used in the construction of the cavity is class B-s3, d2 or better.

   ii. The supports and fixings in the cavity are class A1.

e. Any pipe insulation system should achieve class C-s3, d2 rating or better.

f. Any electrical wiring in the cavity is laid in metal trays or metal conduit.

g. Other than those in (d)–(f), any materials are class A2-s3, d2 rated or better.

Construction and fixings for cavity barriers

9.13 Cavity barriers, tested from each side separately, should provide a minimum of both of the following:

a. 30 minutes' integrity (E 30)

b. 15 minutes' insulation (I 15).

They may be formed by a construction provided for another purpose if it achieves the same performance.

9.14 Cavity barriers in a stud wall or partition, or provided around openings, may be formed of any of the following.

a. Steel, a minimum of 0.5mm thick.

b. Timber, a minimum of 38mm thick.

c. Polythene-sleeved mineral wool, or mineral wool slab, under compression when installed in the cavity.

d. Calcium silicate, cement-based or gypsum-based boards, a minimum of 12mm thick.

These do not necessarily achieve the performance specified in paragraph 9.13.

NOTE: Cavity barriers provided around openings may be formed by the window or door frame if the frame is constructed of steel or timber of the minimum thickness in (a) or (b), as appropriate.

9.15 Cavity barriers should be tightly fitted to a rigid construction and mechanically fixed in position. If this is not possible (e.g. where a cavity barrier joins to slates, tiles, corrugated sheeting or similar materials) the junction should be fire-stopped.

9.16 Cavity barriers should be fixed so their performance is unlikely to be made ineffective by any of the following.

a. Movement of the building due to subsidence, shrinkage or temperature change, and movement of the external envelope due to wind.

b. During a fire, collapse of services penetrating the cavity barriers, either by the failure of the supporting system or through degradation of the service itself (e.g. by melting or burning).

c. During a fire, failure of the cavity barrier fixings. (In roof spaces, where cavity barriers are fixed to roof members, there is no expectation of fire resistance from roof members provided for the purpose of support.)

d. During a fire, failure of any material or construction to which cavity barriers abut. (For example, a suspended ceiling that continues over a fire resisting wall or partition collapses, and the cavity barrier fails prematurely because the ceiling was not designed to provide a minimum fire resistance of EI 30.)

Openings in cavity barriers

9.17 Openings should be limited to the following.

a. Fire doorsets with a minimum rating of E 30, fitted in accordance with Appendix C.

b. The passage of pipes that follow the provisions in Section 10.

c. The passage of cables or conduits containing one or more cables.

d. Openings fitted with a suitably mounted and appropriate fire damper.

e. Ducts that are either of the following.

   i. Fire resisting (minimum E 30).

   ii. Fitted with a suitably mounted and appropriate fire damper where they pass through the cavity barrier.

9.18 If a cavity barrier is provided above or below a partition between bedrooms in 'residential (institutional)' and 'residential (other)' (purpose groups 2(a) and 2(b)) buildings, and the partition is not a fire resisting partition, then paragraph 9.17 does not apply. However, both of the following apply.

a. The number of openings in the barrier should be kept to a minimum.

b. Any penetrations should be sealed to restrict the passage of smoke with an appropriate fire-stopping material.

NOTE: For further guidance on openings in cavity barriers see Section 10.

Section 10: Protection of openings and fire-stopping

Introduction

10.1 The performance of a fire-separating element should not be impaired. Every joint, imperfect fit and opening for services should be sealed. Fire-stopping delays the spread of fire and, generally, the spread of smoke as well.

Openings for pipes

10.2 Pipes passing through a fire-separating element, unless in a protected shaft, should comply with one of the alternatives A, B or C below.

Alternative A: Proprietary seals (any pipe diameter)

10.3 Provide a proprietary, tested sealing system that will maintain the fire resistance of the wall, floor or cavity barrier.

Alternative B: Pipes with a restricted diameter

10.4 Where a proprietary sealing system is not used, fire-stop around the pipe, keeping the opening for the pipe as small as possible. The nominal internal diameter of the pipe should not exceed the relevant dimension given in Table 10.1.

Alternative C: Sleeving

10.5 A pipe with a maximum nominal internal diameter of 160mm may be used with a sleeve made out of a high melting point metal, as shown in Diagram 10.1, if the pipe is made of one of the following.

a. Lead.

b. Aluminium.

c. Aluminium alloy.

d. Fibre-cement.

e. uPVC (pipes should also comply with either **BS 4514** or **BS 5255**).

A high melting point metal means any metal (such as cast iron, copper or steel) which, if exposed to a temperature of 800°C, will not soften or fracture to the extent that flame or hot gas will pass through the wall of the pipe.

See para 10.5

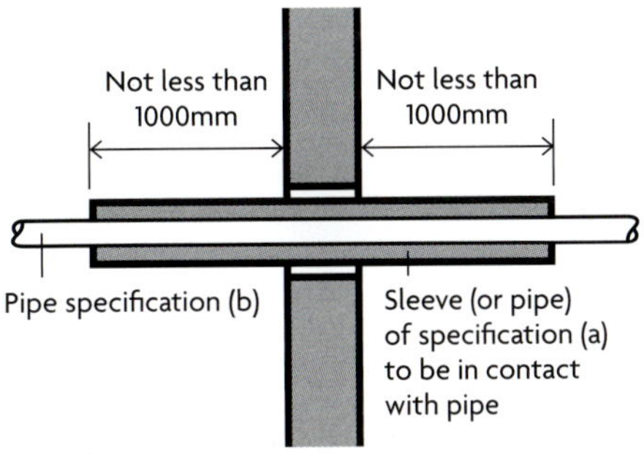

NOTES:

1. Make the opening in the structure as small as possible and provide fire-stopping between pipe and structure.
2. See Table 10.1 for materials specification.
3. The sleeve should be class A1 rated.

Diagram 10.1 Pipes penetrating structure

Table 10.1 Maximum nominal internal diameter of pipes passing through a compartment wall/floor

Situation	Pipe material and maximum nominal internal diameter (mm)		
	(a) High melting point metal[1]	(b) Lead, aluminium, aluminium alloy, uPVC[2], fibre-cement	(c) Any other material
1. Structure (but not a wall separating buildings) enclosing a protected shaft that is not a stairway or a lift shaft	160	110	40
2. Any other situation	160	40	40

NOTES:

1. Any metal (such as cast iron, copper or steel) which, if exposed to a temperature of 800°C, will not soften or fracture to the extent that flame or hot gas will pass through the wall of the pipe.
2. uPVC pipes that comply with either **BS 4514** or **BS 5255**.

Mechanical ventilation and air-conditioning systems

General provisions

10.6 Ductwork should not help to transfer fire and smoke through the building. Terminals of exhaust points should be sited away from final exits, cladding or roofing materials achieving class B-s3, d2 or worse and openings into the building.

10.7 Ventilation ducts supplying or extracting air directly to or from a protected stairway should not also serve other areas. A separate ventilation system should be provided for each protected stairway.

10.8 A fire and smoke damper should be provided where ductwork enters or leaves each section of the protected escape route it serves. It should be operated by a smoke detector or suitable fire detection system. Fire and smoke dampers should close when smoke is detected. Alternatively, the methods set out in paragraphs 10.15 and 10.16 and Diagrams 10.2 and 10.3 may be followed.

10.9 In a system that recirculates air, smoke detectors should be fitted in the extract ductwork before both of the following.

a. The point where recirculated air is separated from air to be discharged to the outside.

b. Any filters or other air cleaning equipment.

When smoke is detected, detectors should do one of the following.

a. Cause the system to immediately shut down.

b. Switch the ventilation system from recirculating mode to extraction to divert smoke outside the building.

10.10 Non-domestic kitchens, car parks and plant rooms should have separate and independent extraction systems. Extracted air should not be recirculated.

10.11 Under fire conditions, ventilation and air-conditioning systems should be compatible with smoke control systems and need to be considered in their respective design.

Ventilation ducts and flues passing through fire-separating elements

General provisions

10.12 If air handling ducts pass through fire-separating elements, the load-bearing capacity, integrity and insulation of the elements should be maintained using one or more of the following four methods. In most ductwork systems, a combination of the four methods is best.

a. Method 1 – thermally activated fire dampers.

b. Method 2 – fire resisting enclosures.

c. Method 3 – protection using fire resisting ductwork.

d. Method 4 – automatically activated fire and smoke dampers triggered by smoke detectors.

10.13 Further information on fire resisting ductwork is given in the ASFP Blue Book.

Kitchen extract

10.14 Methods 1 and 4 should not be used for extract ductwork serving kitchens. The likely build-up of grease within the duct can adversely affect dampers.

Ducts passing through protected escape routes

10.15 Method 1 should not be used for extract ductwork passing through the enclosures of protected escape routes (Diagrams 10.2 and 10.3), as large volumes of smoke can pass thermal devices without triggering them.

10.16 An ES classified fire and smoke damper which is activated by a suitable fire detection system (method 4) may also be used for protected escape routes.

See para 10.15

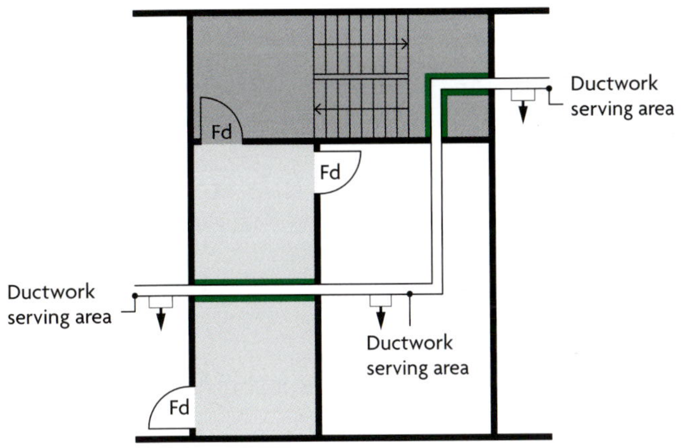

- Ductwork enclosed in fire resisting construction classified EI X in accordance with BS EN 13501-2 (fire exposure from the duct side), or fire resisting ductwork classified EIS X in accordance with BS EN 13501-3, where X is the fire resistance rating (in minutes) of the walls of the protected escape route
- Protected stairway
- Protected lobby
- Fd Fire doorset

NOTE: Ventilation ducts which serve other parts of the building should not supply or extract air directly to or from a protected escape route.

Diagram 10.2 Ductwork passing through protected escape routes – method 2 or method 3

See para 10.15

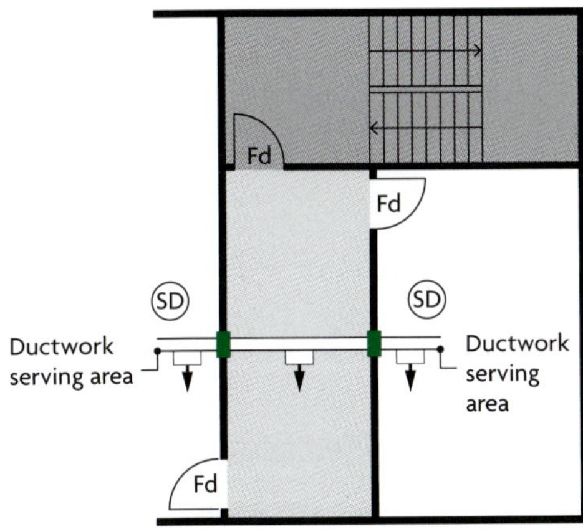

- ES leakage rated fire and smoke damper conforming to BS EN 13501-3/BS EN 1366-2
- Protected stairway
- Protected lobby
- (SD) Smoke detection system in accordance with BS 5839-1 to activate ES damper
- Fd Fire doorset

NOTE: Ventilation ducts which serve other parts of the building should not supply or extract air directly to or from a protected escape route.

Diagram 10.3 Ductwork passing through protected escape routes – method 4

Installation and specification of fire dampers

10.17 Both fire dampers and fire and smoke dampers should be all of the following.

a. Sited within the thickness of the fire-separating elements.

b. Securely fixed.

c. Sited such that, in a fire, expansion of the ductwork would not push the fire damper through the structure.

10.18 Access to the fire damper and its actuating mechanism should be provided for inspection, testing and maintenance.

10.19 Fire dampers should meet both of the following conditions.

a. Conform to **BS EN 15650**.

b. Have a minimum E classification of 60 minutes or to match the integrity rating of the fire resisting elements, whichever is higher.

10.20 Fire and smoke dampers should meet both of the following conditions.

a. Conform to **BS EN 15650**.

b. Have a minimum ES classification of 60 minutes or to match the integrity rating of the fire resisting elements, whichever is higher.

10.21 Smoke detectors should be sited so as to prevent the spread of smoke as early as practicable by activating the fire and smoke dampers. Smoke detectors and automatic release mechanisms used to activate fire dampers and/or fire and smoke dampers should conform to **BS EN 54-7** and **BS 5839-3** respectively.

Further information on fire dampers and/or fire and smoke dampers is given in the ASFP Grey Book.

Sleeping risks

10.22 Where the use of the building involves a sleeping risk, fire dampers or fire and smoke dampers should be actuated by both of the following.

a. Smoke detector-controlled automatic release mechanisms.

b. Thermally actuated devices.

However, in a situation where both of the following are true:

a. all occupants of the building can be expected to make an unaided escape

b. an L1 fire detection and alarm system is installed in accordance with **BS 5839-1**

then both of the following exceptions may be made.

i. If, on the detection of smoke, the fire alarm system signals the immediate evacuation of all the occupants of the building, then fire dampers and/or fire and smoke dampers do not need to be actuated by smoke detectors.

ii. If the building is divided into fire compartments and the alarm system is arranged to signal the immediate evacuation of the occupants of the fire compartment in which the fire has been detected, then smoke detector-operated fire dampers or fire and smoke dampers need only be provided where ductwork enters or leaves the fire compartment.

Flues, etc.

10.23 The wall of a flue, duct containing flues or appliance ventilation duct(s) should have a fire resistance (REI) that is at least half of any compartment wall or compartment floor it passes through or is built into (Diagram 10.4).

See para 10.23

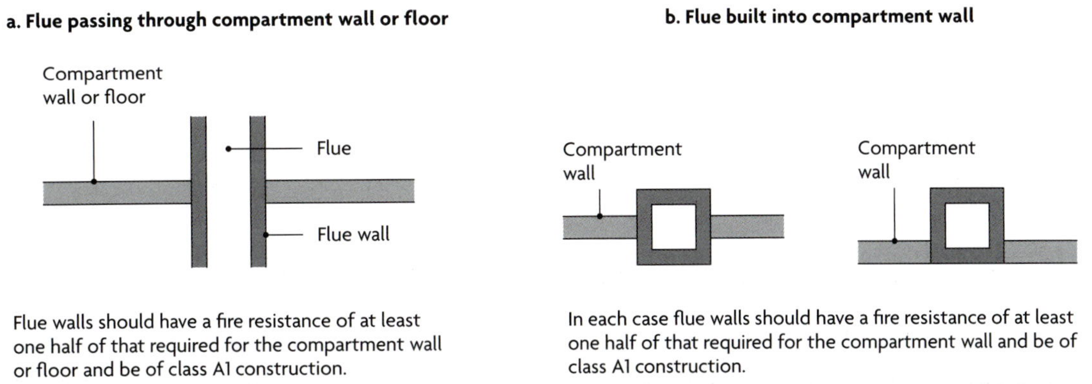

Diagram 10.4 Flues penetrating compartment walls or floors

Fire-stopping

10.24 In addition to any other provisions in this section, both of the following conditions should be met.

a. Joints between fire-separating elements should be fire-stopped.

b. Openings through a fire resisting element for pipes, ducts, conduits or cable should be all of the following.

  i. As few as possible.

  ii. As small as practicable.

  iii. Fire-stopped (allowing thermal movement in the case of a pipe or duct).

NOTE: The fire-stopping around fire dampers, fire resisting ducts, fire and smoke dampers and smoke control ducts should be in accordance with the manufacturer or supplier's installation instructions.

10.25 Materials used for fire-stopping should be reinforced with (or supported by) materials rated class A2-s3, d2 or better to prevent displacement in both of the following cases.

a. Where the unsupported span is greater than 100mm.

b. Where non-rigid materials are used (unless subjected to appropriate fire resistance testing to show their suitability).

10.26 Proprietary, tested fire-stopping and sealing systems are available and may be used. Different materials suit different situations and not all are suitable in every situation.

10.27 Other fire-stopping materials include the following.

a. Cement mortar.

b. Gypsum-based plaster.

c. Cement-based or gypsum-based vermiculite/perlite mixes.

d. Glass fibre, crushed rock, blast furnace slag or ceramic-based products (with or without resin binders).

e. Intumescent mastics.

These may be used in situations appropriate to the particular material. Not all materials will be suitable in every situation.

10.28 Guidance on the design, installation and maintenance of measures to contain fires or slow their spread is given in *Ensuring Best Practice for Passive Fire Protection in Buildings* produced by the Association for Specialist Fire Protection (ASFP).

10.29 Further information on generic systems, their suitability for different applications and guidance on test methods is given in the ASFP Red Book.

Section 11: Special provisions for car parks

11.1 Car parks call for different measures to restrict fire spread within buildings for the following reasons.

a. The fire load is well defined.

b. The probability of fire spreading from one storey to another in a well ventilated car park is low. Guidance is therefore given for three ventilation scenarios.

Open-sided car parks

Natural ventilation

11.2 For the purposes of assessing fire resistance, a building, compartment or separated part containing a car park may be regarded as open-sided when it complies with all of the following.

a. There are no basement storeys.

b. Each storey is naturally ventilated by permanent openings at each car parking level. The aggregate vent area is a minimum of 1/20 of that level's floor area, at least half of which is provided equally by two opposite walls.

c. Where one element of structure supports, carries or stabilises another, the fire resistance of the supporting element at least matches the minimum period of fire resistance for the other element.

d. In mixed use buildings, the fire resistance of any element that supports, carries or stabilises an element in the other part of the building should at least match the minimum period of fire resistance for the other element.

e. All materials used in the construction should be class A1 rated, except for the following.

i. Any surface finish applied to a floor or roof of the car park (or within any building, compartment or separated part adjoining the structure enclosing it), if the finish meets requirements B2 and B4.

ii. Any fire doorset.

iii. Any attendant's kiosk not exceeding $15m^2$ in area.

iv. Any shop mobility facility.

Car parks that are not open-sided

11.3 For car parks that do not have the ventilation set out in paragraph 11.2, the required fire resistance is given in Appendix B, Table B4. Ventilation should be either natural or mechanical. See Approved Document F for additional guidance on ventilation of car parks.

Natural ventilation

11.4 Each storey should be ventilated by permanent openings at each car parking level. The openings can be at ceiling level. The aggregate free vent area should be a minimum of 1/40 of that level's floor area, at least half of which should be provided equally by two opposite walls (1/160 on each side). The remaining free area can be distributed wherever possible.

Mechanical ventilation

11.5 If the minimum standard of natural ventilation is not possible, a system of mechanical ventilation should be provided that complies with all of the following.

a. The system should be both of the following.

   i. Independent of any other ventilating system (other than any system that provides day to day ventilation to the car park).

   ii. Designed to operate at 10 air changes per hour during a fire.

b. The system should run in two parts, each of which is:

   i. capable of extracting 50% of the rates set out in item (a)

   ii. able to operate alone or with the other part

   iii. provided with an independent power supply capable of operating if the main supply fails.

c. 50% of the outlets should be at high level and 50% at low level.

d. The system should use E, I and S ductwork in accordance with **BS EN 1366-8**.

For further information on equipment for removing hot smoke, refer to **BS EN 12101-3**. An alternative method of providing smoke ventilation from enclosed car parks is given in **BS 7346-7**.

Requirement B4: External fire spread

These sections deal with the following requirement from Part B of Schedule 1 to the Building Regulations 2010. Section 12 also refers to regulation 7(2) of the Building Regulations 2010. Guidance on regulation 7(1) can be found in Approved Document 7.

Requirement

Requirement	*Limits on application*
External fire spread **B4.** (1) The external walls of the building shall adequately resist the spread of fire over the walls and from one building to another, having regard to the height, use and position of the building. (2) The roof of the building shall adequately resist the spread of fire over the roof and from one building to another, having regard to the use and position of the building.	

Regulation

Regulation 7 – Materials and workmanship

(1) Building work shall be carried out—

(a) with adequate and proper materials which—

(i) are appropriate for the circumstances in which they are used,

(ii) are adequately mixed or prepared, and

(iii) are applied, used or fixed so as adequately to perform the functions for which they are designed; and

(b) in a workmanlike manner.

(2) Subject to paragraph (3), building work shall be carried out so that materials which become part of an external wall, or specified attachment, of a relevant building are of European Classification A2-s1, d0 or Class A1, classified in accordance with BS EN 13501-1:2007+A1:2009 entitled "Fire classification of construction products and building elements. Classification using test data from reaction to fire tests" (ISBN 978 0 580 59861 6) published by the British Standards Institution on 30th March 2007 and amended in November 2009.

Regulation *continued*

(3) Paragraph (2) does not apply to—

(a) cavity trays when used between two leaves of masonry;

(b) any part of a roof (other than any part of a roof which falls within paragraph (iv) of regulation 2(6)) if that part is connected to an external wall;

(c) door frames and doors;

(d) electrical installations;

(e) insulation and water proofing materials used below ground level;

(f) intumescent and fire stopping materials where the inclusion of the materials is necessary to meet the requirements of Part B of Schedule 1;

(g) membranes;

(h) seals, gaskets, fixings, sealants and backer rods;

(i) thermal break materials where the inclusion of the materials is necessary to meet the thermal bridging requirements of Part L of Schedule 1; or

(j) window frames and glass.

(4) In this regulation—

(a) a "relevant building" means a building with a storey (not including roof-top plant areas or any storey consisting exclusively of plant rooms) at least 18 metres above ground level and which—

(i) contains one or more dwellings;

(ii) contains an institution; or

(iii) contains a room for residential purposes (excluding any room in a hostel, hotel or boarding house); and

(b) "above ground level" in relation to a storey means above ground level when measured from the lowest ground level adjoining the outside of a building to the top of the floor surface of the storey.

Intention

Resisting fire spread over external walls

The external envelope of a building should not contribute to undue fire spread from one part of a building to another part. This intention can be met by constructing external walls so that both of the following are satisfied.

a. The risk of ignition by an external source to the outside surface of the building and spread of fire over the outside surface is restricted.

b. The materials used to construct external walls, and attachments to them, and how they are assembled do not contribute to the rate of fire spread up the outside of the building.

The extent to which this is necessary depends on the height and use of the building.

Resisting fire spread from one building to another

The external envelope of a building should not provide a medium for undue fire spread to adjacent buildings or be readily ignited by fires in adjacent buildings. This intention can be met by constructing external walls so that all of the following are satisfied.

a. The risk of ignition by an external source to the outside surface of the building is restricted.

b. The amount of thermal radiation that falls on a neighbouring building from window openings and other unprotected areas in the building on fire is not enough to start a fire in the other building.

c. Flame spread over the roof and/or fire penetration from external sources through the roof is restricted.

The extent to which this is necessary depends on the use of the building and its position in relation to adjacent buildings and therefore the site boundary.

Section 12: Resisting fire spread over external walls

Introduction

12.1 The external wall of a building should not provide a medium for fire spread if that is likely to be a risk to health and safety. Combustible materials and cavities in external walls and attachments to them can present such a risk, particularly in tall buildings. The guidance in this section is designed to reduce the risk of vertical fire spread as well as the risk of ignition from flames coming from adjacent buildings.

Fire resistance

12.2 This section does not deal with fire resistance for external walls. An external wall may need fire resistance to meet the requirements of Section 5 (General provisions), Section 7 (Loadbearing elements of structures) or Section 13 (Resisting fire spread from one building to another).

Combustibility of external walls

12.3 The external walls of buildings other than those described in regulation 7(4) of the Building Regulations should achieve either of the following.

a. Follow the provisions given in paragraphs 12.5 to 12.9, which provide guidance on all of the following.

   i. External surfaces.

   ii. Materials and products.

   iii. Cavities and cavity barriers.

b. Meet the performance criteria given in BRE report BR 135 for external walls using full-scale test data from **BS 8414-1** or **BS 8414-2**.

12.4 In relation to buildings of any height or use, consideration should be given to the choice of materials (including their extent and arrangement) used for the external wall, or attachments to the wall, to reduce the risk of fire spread over the wall.

External surfaces

12.5 The external surfaces (i.e. outermost external material) of external walls should comply with the provisions in Table 12.1. The provisions in Table 12.1 apply to each wall individually in relation to its proximity to the relevant boundary.

Table 12.1 Reaction to fire performance of external surface of walls

Building type	Building height	Less than 1000mm from the relevant boundary	1000mm or more from the relevant boundary
'Relevant buildings' as defined in regulation 7(4) (see paragraph 12.11)		Class A2-s1, d0[1] or better	Class A2-s1, d0[1] or better
Assembly and recreation	More than 18m	Class B-s3, d2[2] or better	From ground level to 18m: class C-s3, d2[3] or better From 18m in height and above: class B-s3, d2[2] or better
	18m or less	Class B-s3, d2[2] or better	Up to 10m above ground level: class C-s3, d2[3] or better Up to 10m above a roof or any part of the building to which the public have access: class C-s3, d2[3] or better[4] From 10m in height and above: no minimum performance
Any other building	More than 18m	Class B-s3, d2[2] or better	From ground level to 18m: class C-s3, d2[3] or better From 18m in height and above: class B-s3, d2[2] or better
	18m or less	Class B-s3, d2[2] or better	No provisions

NOTES:

In addition to the requirements within this table, buildings with a top occupied storey above 18m should also meet the provisions of paragraph 12.6.

In all cases, the advice in paragraph 12.4 should be followed.

1. The restrictions for these buildings apply to all the materials used in the external wall and specified attachments (see paragraphs 12.10 to 12.13 for further guidance).
2. Profiled or flat steel sheet at least 0.5 mm thick with an organic coating of no more than 0.2mm thickness is also acceptable.
3. Timber cladding at least 9mm thick is also acceptable.
4. 10m is measured from the top surface of the roof.

Materials and products

12.6 In a building with a storey 18m or more in height (see Diagram D6 in Appendix D) any insulation product, filler material (such as the core materials of metal composite panels, sandwich panels and window spandrel panels but not including gaskets, sealants and similar) etc. used in the construction of an external wall should be class A2-s3, d2 or better (see Appendix B). This restriction does not apply to masonry cavity wall construction which complies with Diagram 9.2 in Section 9. Where regulation 7(2) applies, that regulation prevails over all the provisions in this paragraph.

12.7 Best practice guidance for green walls (also called living walls) can be found in *Fire Performance of Green Roofs and Walls*, published by the Department for Communities and Local Government.

Cavities and cavity barriers

12.8 Cavity barriers should be provided in accordance with Section 9.

12.9 In the case of an external wall construction of a building which, by virtue of paragraph 9.10d (external cladding system with a masonry or concrete inner leaf), is not subject to the provisions of Table 9.1, the surfaces which face into cavities should also meet the provisions of Table 12.1 and provisions in Section 9, but where regulation 7(2) applies, that regulation prevails over the guidance provided in Table 12.1 and Section 9.

Regulation 7(2) and requirement B4

Materials

12.10 Regulation 7(1)(a) requires that materials used in building work are appropriate for the circumstances in which they are used. Regulation 7(2) sets requirements in respect of external walls and specified attachments in relevant buildings.

NOTE: Guidance on regulation 7(1) can be found in Approved Document 7.

12.11 Regulation 7(2) applies to any building with a storey at least 18m above ground level (as measured in accordance with Diagram D6 in Appendix D) and which contains one or more dwellings; an institution; or a room for residential purposes (excluding any room in a hostel, hotel or a boarding house). It requires that all materials which become part of an external wall or specified attachment achieve class A2-s1, d0 or class A1, other than those exempted by regulation 7(3).

NOTE: The above includes student accommodation, care homes, sheltered housing, hospitals and dormitories in boarding schools. See regulation 7(4) for the definition of relevant buildings.

NOTE: The requirement in regulation 7(2) is limited to materials achieving class A2-s1, d0 or class A1.

12.12 External walls and specified attachments are defined in regulation 2 and these definitions include any parts of the external wall as well as balconies, solar panels and sun shading.

12.13 Regulation 7(3) provides an exemption for certain components found in external walls and specified attachments.

Material change of use

12.14 Regulations 5(k) and 6(3) provide that, where the use of a building is changed such that the building becomes a building described in regulation 7(4), the construction of the external walls, and specified attachments, must be investigated and, where necessary, work must be carried out to ensure they only contain materials achieving class A2-s1, d0 or class A1, other than those exempted by regulation 7(3).

Additional considerations

12.15 The provisions of regulation 7 apply in addition to requirement B4. Therefore, for buildings described in regulation 7(4), the potential impact of any products incorporated into or onto the external walls and specified attachments should be carefully considered with regard to their number, size, orientation and position.

12.16 Particular attention is drawn to the following points.

a. Membranes used as part of the external wall construction above ground level should achieve a minimum of class B-s3, d0.

b. Internal linings should comply with the guidance provided in Section 6.

c. Any part of a roof should achieve the minimum performance as detailed in Section 14.

d. As per regulation 7(3), window frames and glass (including laminated glass) are exempted from regulation 7(2). Window spandrel panels and infill panels must comply with regulation 7(2).

e. Thermal breaks are small elements used as part of the external wall construction to restrict thermal bridging. There is no minimum performance for these materials. However, they should not span two compartments and should be limited in size to the minimum required to restrict the thermal bridging (the principal insulation layer is not to be regarded as a thermal break).

f. Regulation 7(2) only applies to specified attachments. Shop front signs and similar attachments are not covered by the requirements of regulation 7(2), although attention is drawn to paragraph 12.16g.

g. While regulation 7(2) applies to materials which become part of an external wall or specified attachment, consideration should be given to other attachments to the wall which could impact on the risk of fire spread over the wall.

Section 13: Resisting fire spread from one building to another

Introduction

13.1 The following assumptions enable a reasonable standard of resistance to the spread of fire to be specified.

a. The size of a fire depends on the compartmentation within the building. A fire may involve a complete compartment, but will not spread to other compartments.

b. The intensity of fire is related to the building use, but can be moderated by a sprinkler system.

c. Fires in 'residential' and 'assembly and recreation' buildings (purposes groups 1, 2 and 5) represent a greater risk to life.

d. A building on the far side of the relevant boundary meets both of the following conditions.

i. Has a similar elevation to the one in question.

ii. Is at the same distance from the common boundary.

e. The radiated heat passing through any part of the fire resisting external wall may be discounted.

13.2 Where regulation 7(2) applies, that regulation prevails over the provisions within this section.

13.3 If a reduced separation distance between buildings, or increased amount of unprotected area, is required, smaller compartments should be considered.

Boundaries

13.4 The fire resistance of a wall depends on its distance from the relevant boundary (see Diagram 13.1). Separation distances are measured to boundaries to ensure that the location and design of buildings on adjoining sites have no influence on the building under consideration.

13.5 The boundary that a wall faces is the relevant boundary (Diagram 13.2). It may be one of the following.

a. The site boundary.

b. The centre line of a space where further development is unlikely, such as a road, railway, canal or river.

c. An assumed notional boundary between two buildings on the same site (Diagram 13.3) where either of the following conditions is met.

i. One or both of the buildings are in the 'residential' or 'assembly and recreation' purpose groups (purpose group 1 or 5).

ii. The buildings will be operated/managed by different organisations.

See para 13.4

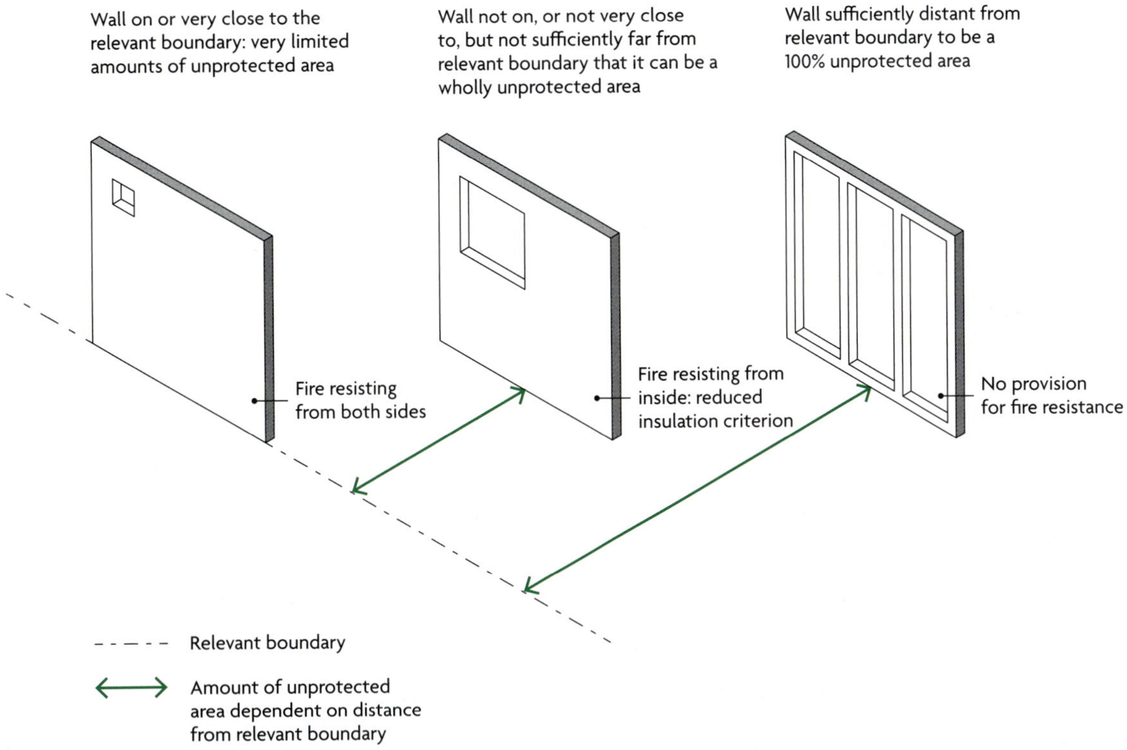

Diagram 13.1 Principles of space separation

See para 13.5

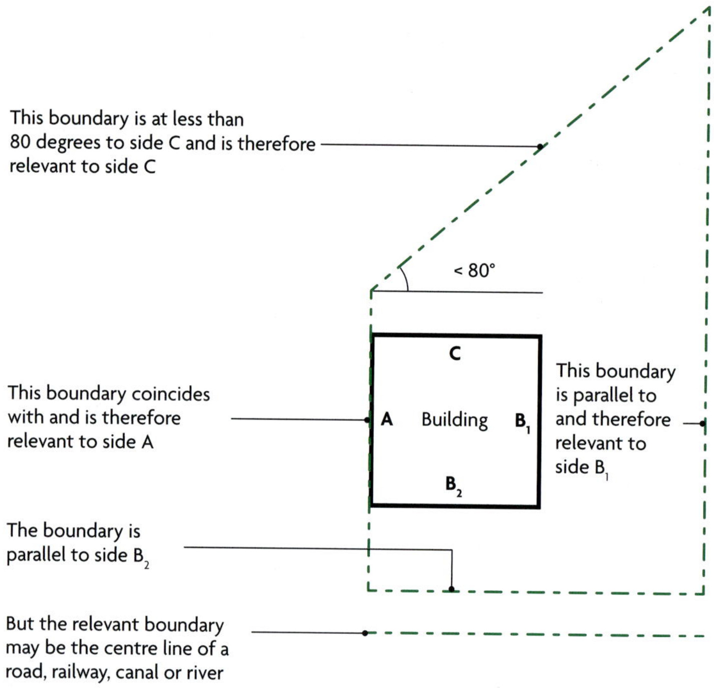

NOTES:

This diagram sets out the rules that apply in respect of a boundary for it to be considered as a relevant boundary.

For a boundary to be relevant it should comply with one of the following:

a. Coincide with the side of the building (**A**).
b. Be parallel to the side of the building (**B_1** or **B_2**).
c. Be at an angle of maximum 80 degrees to the side of the building (**C**).

Diagram 13.2 Relevant boundary

See para 13.5

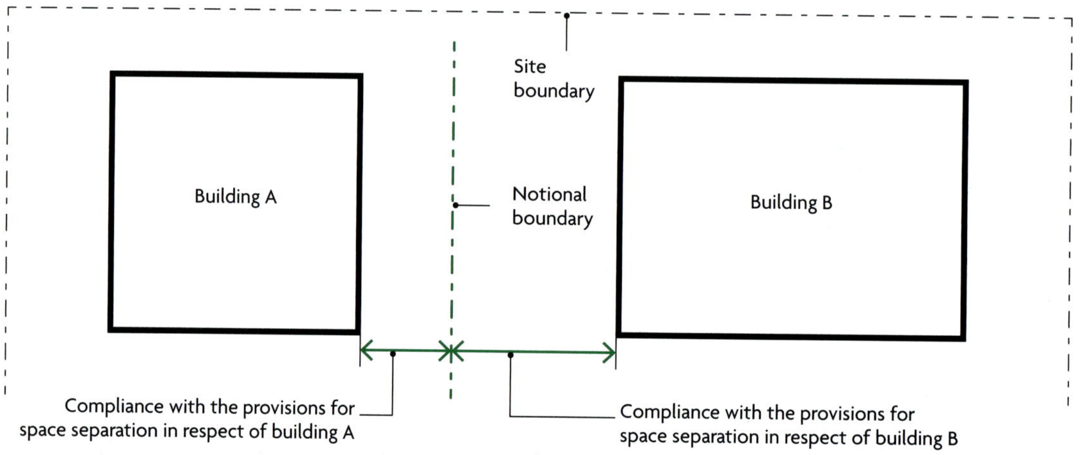

NOTES:

The notional boundary should be set in the area between the two buildings using the following rules:

1. The notional boundary is assumed to exist in the space between the buildings and is positioned so that one of the buildings would comply with the provisions for space separation having regard to the amount of its unprotected area. In practice, if one of the buildings is existing, the position of the boundary will be set by the space separation factors for that building.
2. The siting of the new building, or the second building if both are new, can then be checked to see that it also complies, using the notional boundary as the relevant boundary for the second building.

Diagram 13.3 Notional boundary

Unprotected areas and fire resistance

13.6 Parts of an external wall with less fire resistance than the appropriate amount given in Appendix B, Table B4, are called unprotected areas.

13.7 Where a fire resisting external wall has a surface material that is worse than class B-s3, d2 and is more than 1mm thick, that part of the wall should be classified as an unprotected area equating to half its area (Diagram 13.4).

External walls on, and within 1000mm of, the relevant boundary

13.8 Unprotected areas should meet the conditions in Diagram 13.5 and the rest of the wall should be fire resisting from both sides.

External surface materials facing the boundary should be class B-s3, d2 or better.

External walls 1000mm or more from the relevant boundary

13.9 Unprotected areas should not exceed the result given by one of the methods in paragraph 13.17, and the rest of the wall (if any) should be fire resisting but only from the inside of the building.

External walls of protected stairways

13.10 Exclude external walls of protected stairways when assessing unprotected areas (see Diagram 3.3).

See para 13.7

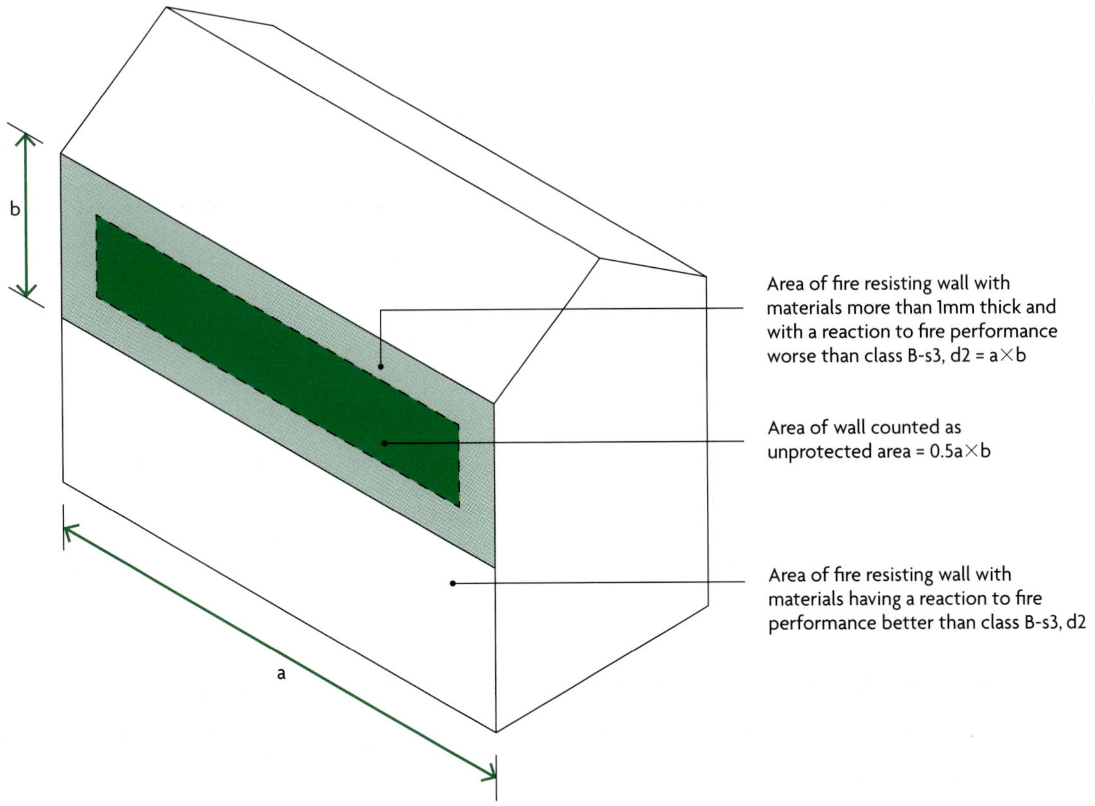

Diagram 13.4 Status of materials achieving class B-s3, d2 or worse as unprotected area

Small unprotected areas

13.11 In an otherwise protected wall, small unprotected areas may be ignored where they meet the conditions in Diagram 13.5.

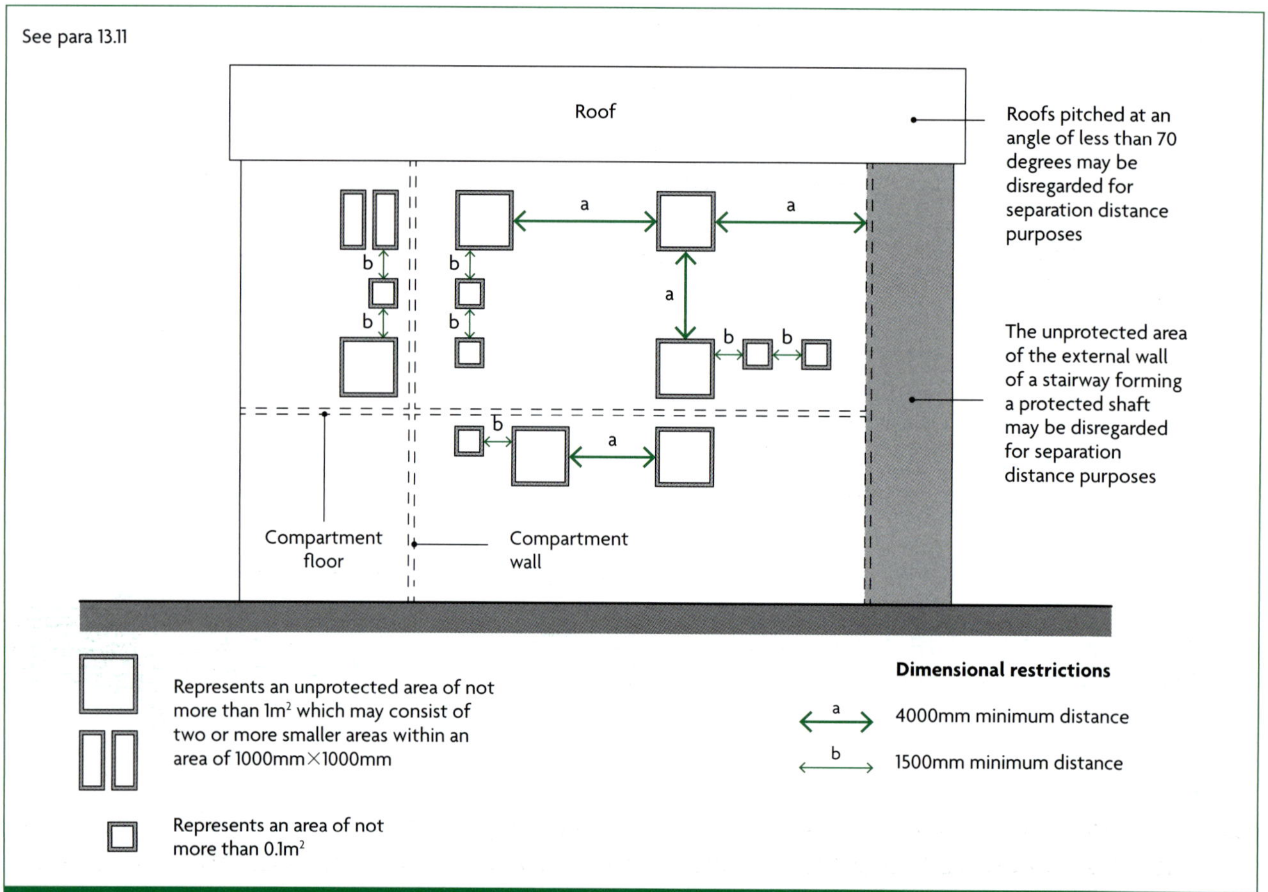

Diagram 13.5 Small unprotected areas that may be disregarded in assessing the separation distance from the boundary

Large uncompartmented buildings

13.12 For the purposes of assessing unprotected area, parts of walls of uncompartmented buildings that are more than 30m above mean ground level may be ignored.

Canopies

13.13 Where both of the following apply, separation distances may be determined from the wall rather than from the edge of the canopy (Diagram 13.6).

a. The canopy is attached to the side of a building.

b. The edges of the canopy are a minimum of 2m from the relevant boundary.

Canopies that fall within class 6 or class 7 of Schedule 2 to the regulations (Exempt Buildings and Work) are exempt from the Building Regulations.

13.14 Space separation may be disregarded if a canopy is all of the following.

a. Free-standing.

b. Above a limited risk or controlled hazard, for example over petrol pumps.

c. A minimum of 1000mm from the relevant boundary.

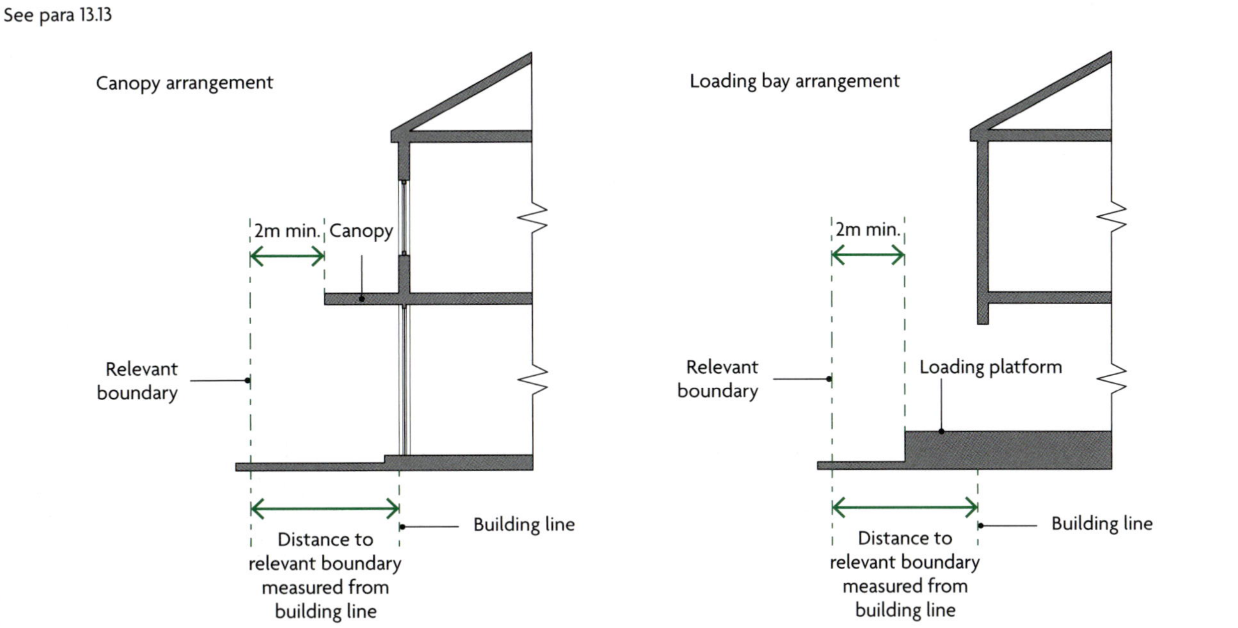

NOTE: Projections from the building line, such as a canopy or a loading platform, can be ignored when assessing separation distance. This does not apply where the canopy is enclosed by side walls.

Diagram 13.6 The effect of a canopy on separation distance

Roofs

13.15 Roofs with a pitch of more than 70 degrees to the horizontal should be assessed in accordance with this section. Vertical parts of a pitched roof, such as dormer windows, should be included, *only* if the slope of the roof exceeds 70 degrees.

It is a matter of judgement whether a continuous run of dormer windows that occupies most of a steeply pitched roof should be treated as a wall rather than a roof.

Portal frames

13.16 Portal frames are often used in single storey industrial and commercial buildings where there may be no need for fire resistance of the structure (requirement B3). However, where a portal framed building is near a relevant boundary, the external wall near the boundary may need fire resistance to restrict the spread of fire between buildings. It is generally accepted that a portal frame acts as a single structural element because of the moment-resisting connections used, especially at the column/rafter joints. Thus, in cases where the external wall of the building cannot be wholly unprotected, the rafter members of the frame, as well as the column members, may need to be fire protected. The design method for this is set out in SCI Publication P313.

NOTE: The recommendations in the SCI publication for designing the foundation to resist overturning do not need to be followed if the building is fitted with a sprinkler system in accordance with Appendix E.

NOTE: Normally, portal frames of reinforced concrete can support external walls requiring a similar degree of fire resistance without specific provision at the base to resist overturning.

NOTE: Existing buildings may have been designed to comply with all of the following guidance, which is also acceptable.

a. The column members are fixed rigidly to a base of sufficient size and depth to resist overturning.

b. There is brick, block or concrete protection to the columns up to a protected ring beam providing lateral support.

c. There is some form of roof venting to give early heat release. (The roof venting could be, for example, PVC rooflights covering some 10% of the floor area and evenly spaced over the floor area.)

Methods for calculating acceptable unprotected area

13.17 Two simple methods are given for calculating the acceptable amount of unprotected area in an external wall that is a minimum of 1000mm from any point on the relevant boundary. More precise methods are described in BRE report BR 187 and may be used instead.

Method 1

13.18 This method applies to small buildings intended to be used for 'residential (other)' purposes.

13.19 The building should not exceed three storeys in height (excluding basements) or 24m in length. Each side of the building should meet the limits stated in Diagram 13.7. Any small unprotected areas falling within the limits shown in Diagram 13.5 can be ignored.

See para 13.19

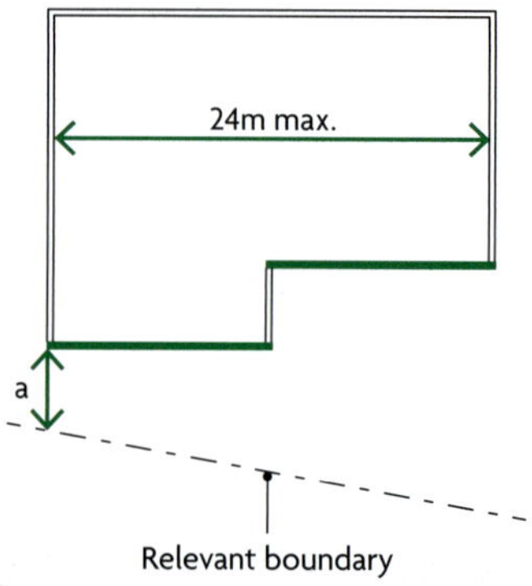

Minimum distance (a) between side of building and relevant boundary (m)	Maximum total area of unprotected areas (m²)
1	5.6
2	12
3	18
4	24
5	30
6	No limit

Diagram 13.7 Permitted unprotected areas in small residential buildings

Method 2

13.20 This method may be used for buildings or compartments intended for any use and for which method 1 is not appropriate.

13.21 Except for an open-sided car park in purpose group 7(b) (see paragraph 11.2), the building should not exceed 10m in height. Each side of the building should meet the limits in Table 13.1. Areas falling within the limits in Diagram 13.5 can be ignored.

Table 13.1 Permitted unprotected areas in small buildings or compartments

Minimum distance between side of building and relevant boundary (m)		Maximum total percentage of unprotected area (%)
Purpose groups		
Residential, office, assembly and recreation	Shop and commercial, industrial, storage and other non-residential	
(1)	(2)	(3)
Not applicable	1	4
1	2	8
2.5	5	20
5	10	40
7.5	15	60
10	20	80
12.5	25	100

NOTES:

1. Intermediate values may be obtained by interpolation.
2. For buildings fitted with an automatic sprinkler system, see paragraph 13.22.
3. For open-sided car parks in purpose group 7(b), the distances set out in column (1) may be used instead of those in column (2).
4. The total percentage of unprotected area is found by dividing the total unprotected area by the area of a rectangle that encloses all the unprotected areas, and multiplying the result by 100.

Sprinkler systems

13.22 If a building is fitted throughout with a sprinkler system in accordance with Appendix E, either of the following is permitted.

a. The boundary distance can be halved, to a minimum distance of 1m.

b. The amount of unprotected area can be doubled.

Atrium buildings

13.23 If a building contains one or more atria, the recommendations in clause B8 of **BS 9999** should be followed.

Section 14: Resisting fire spread over roof coverings

Introduction

14.1 'Roof covering' describes one or more layers of material, but not the roof structure as a whole.

14.2 Provisions for the fire properties of roofs are given in other parts of this document.

a. Requirement B1 – for roofs that are part of a means of escape.

b. Requirement B2 – for the internal surfaces of rooflights as part of internal linings.

c. Requirement B3 – for roofs that are used as a floor and for roofs passing over a compartment wall.

d. Section 13 – the circumstances in which a roof is subject to the provisions for space separation.

Separation distances

14.3 Separation distance is the minimum distance from the roof, or part of the roof, to the relevant boundary (paragraph 13.5). Table 14.1 sets out separation distances by the type of roof covering and the size and use of the building.

In addition, roof covering products (and/or materials) defined in Commission Decision 2000/553/EC of 6 September 2000, implementing Council Directive 89/106/EEC, can be considered to fulfil all of the requirements for the performance characteristic 'external fire performance' without the need for testing, *provided that any national provisions on the design and execution of works are fulfilled*, and can be used without restriction.

14.4 The performance of rooflights is specified in a similar way to the performance of roof coverings. Plastic rooflights may also be used.

Plastic rooflights

14.5 Table 14.2 and Diagram 14.1 set the limitations for using plastic rooflights whose lower surface has a minimum rating of class D-s3, d2.

14.6 Table 14.3 sets the limitations for using thermoplastic materials with a TP(a) rigid or TP(b) (see also Diagram 14.1) classification. The method of classifying thermoplastic materials is given in Appendix B.

14.7 Other than for the purposes of Diagram 6.2, polycarbonate or uPVC rooflights achieving a minimum rating of class C-s3, d2 can be regarded as having a B_{ROOF}(t4) classification.

Unwired glass in rooflights

14.8 When used in rooflights, unwired glass a minimum of 4mm thick can be regarded as having a B_{ROOF}(t4) classification.

Thatch and wood shingles

14.9 If the performance of thatch or wood shingles cannot be established, they should be regarded as having an E_{ROOF}(t4) classification in Table 14.1.

See paras 14.5 and 14.6

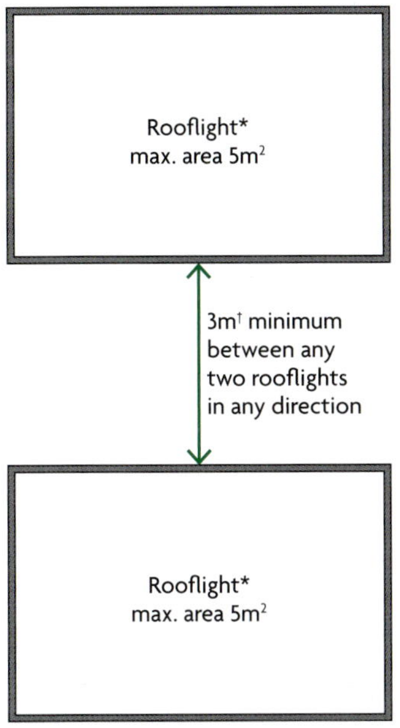

* Or group of rooflights amounting to no more than $5m^2$

† Class D-s3, d2 rooflights to rooms in industrial and other non-residential purpose groups may be spaced 1800mm apart provided the rooflights are evenly distributed and do not exceed 20% of the area of the room

NOTES:

1. There are restrictions on the use of plastic rooflights in the guidance to requirement B2 in Section 6.
2. Surrounding roof covering to be class B-s3, d2 for at least 3m distance.
3. Where Diagram 8.2a or 8.2b applies, rooflights should be at least 1500mm from the compartment wall.

Diagram 14.1 Limitations on spacing and size of plastic rooflights that have a class D-s3, d2 or TP(b) lower surface

Table 14.1 Limitations on roof coverings

Designation[1] of covering of roof or part of roof	Distance from any point on relevant boundary			
	Less than 6m	At least 6m	At least 12m	At least 20m
B_{ROOF}(t4)	●	●	●	●
C_{ROOF}(t4)	○	●	●	●
D_{ROOF}(t4)	○	●[2][3]	●[2]	●
E_{ROOF}(t4)	○	●[2][3]	●[2]	●[2]
F_{ROOF}(t4)	○	○	○	●[2][3]

● Acceptable. ○ Not acceptable.

NOTES:

Separation distances do not apply to enclosed/covered walkways. However, see Diagram 8.2 if the roof passes over the top of a compartment wall.

Polycarbonate and uPVC rooflights that achieve a class C-s3, d2 rating by test may be regarded as having a B_{ROOF}(t4) classification.

1. The designation of external roof surfaces is explained in Appendix B
2. Not acceptable on any of the following buildings.
   a. Industrial, storage or other non-residential purpose group (purpose groups 6 and 7) buildings of any size.
   b. Any other buildings with a cubic capacity of more than $1500m^3$.
3. Acceptable on buildings not listed in (1) if both of the following apply.
   a. Part of the roof has a maximum area of $3m^2$ and is a minimum of 1500mm from any similar part.
   b. The roof between the parts is covered with a material rated class A2-s3, d2 or better.

Table 14.2 Class D-s3, d2 plastic rooflights: limitations on use and boundary distance

Minimum classification on lower surface[1]	Space that rooflight can serve	Minimum distance from any point on relevant boundary to rooflight with an external designation[2] of:	
		E_{ROOF}(t4) or D_{ROOF}(t4)	F_{ROOF}(t4)
Class D-s3, d2	a. Balcony, verandah, carport, covered way or loading bay that has at least one longer side wholly or permanently open b. Detached swimming pool c. Conservatory, garage or outbuilding, with a maximum floor area of $40m^2$	6m	20m
	d. Circulation space[3] (except a protected stairway) e. Room[3]	6m[4]	20m[4]

NOTES:

None of the above designations are suitable for protected stairways.

Polycarbonate and uPVC rooflights that achieve a class C-s3, d2 rating by test (see paragraph 14.7) may be regarded as having a B_{ROOF}(t4) classification.

Where Diagram 8.2a or 8.2b applies, rooflights should be a minimum of 1500m from the compartment wall.

If double-skinned or laminate products have upper and lower surfaces of different materials, the greater distance applies.

1. See also the guidance to requirement B2 in Section 6.
2. The designation of external roof surfaces is explained in Appendix B.
3. Single-skinned rooflight only, in the case of non-thermoplastic material.
4. The rooflight should also meet the provisions of Diagram 14.1.

Table 14.3 TP(a) and TP(b) thermoplastic rooflights: limitations on use and boundary distance

Minimum classification on lower surface[1]	Space that rooflight can serve	Minimum distance from any point on relevant boundary to rooflight with an external surface classification[1] of:	
		TP(a)	TP(b)
1. TP(a) rigid	Any space except a protected stairway	6m[2]	Not applicable
2. TP(b)	a. Balcony, verandah, carport, covered way or loading bay, which has at least one longer side wholly or permanently open b. Detached swimming pool c. Conservatory, garage or outbuilding, with a maximum floor area of $40m^2$	Not applicable	6m
	d. Circulation space[3] (except a protected stairway) e. Room[3]	Not applicable	6m[4]

NOTES:

None of the above designations are suitable for protected stairways.

Polycarbonate and uPVC rooflights that achieve a class C-s3, d2 rating by test may be regarded as having a B_{ROOF}(t4) designation.

Where Diagram 8.2a or 8.2b applies, rooflights should be at least 1500mm from the compartment wall.

If double-skinned or laminate products have upper and lower surfaces of different materials, the greater distance applies.

1. See also the guidance to requirement B2 in Section 6.
2. No limit in the case of any space described in 2(a), (b) and (c).
3. Single-skinned rooflight only, in the case of non-thermoplastic material.
4. The rooflight should also meet the provisions of Diagram 14.1.

Requirement B5: Access and facilities for the fire service

These sections deal with the following requirement from Part B of Schedule 1 to the Building Regulations 2010.

Requirement

Requirement	*Limits on application*
Access and facilities for the fire service **B5.** (1) The building shall be designed and constructed so as to provide reasonable facilities to assist fire fighters in the protection of life. (2) Reasonable provision shall be made within the site of the building to enable fire appliances to gain access to the building.	

Intention

Provisions covering access and facilities for the fire service are to safeguard the health and safety of people in and around the building. Their extent depends on the size and use of the building. Most firefighting is carried out within the building. In the Secretary of State's view, requirement B5 is met by achieving all of the following.

a. External access enabling fire appliances to be used near the building.

b. Access into and within the building for firefighting personnel to both:

   i. search for and rescue people

   ii. fight fire.

c. Provision for internal fire facilities for firefighters to complete their tasks.

d. Ventilation of heat and smoke from a fire in a basement.

If an alternative approach is taken to providing the means of escape, outside the scope of this approved document, additional provisions for firefighting access may be required. Where deviating from the general guidance, it is advisable to seek advice from the fire and rescue service as early as possible (even if there is no statutory duty to consult).

Section 15: Vehicle access

Buildings not fitted with fire mains

15.1 For small buildings (up to 2000m^2, with a top occupied storey that is a maximum of 11m above ground level), vehicle access for a pump appliance should be provided to whichever is the less onerous of the following.

a. 15% of the perimeter.

b. Within 45m of every point of the footprint of the building (see Diagram 15.1).

15.2 For all other buildings, provide vehicle access in accordance with Table 15.1.

15.3 Every elevation to which vehicle access is provided should have a door, a minimum of 750mm wide, to give access into the building. The maximum distance between doors, or between a door and the end of the elevation, is 60m (e.g. a 150m elevation would need a minimum of two doors).

Table 15.1 Fire and rescue service vehicle access to buildings not fitted with fire mains

Total floor area[1] of building (m^2)	Height of floor of top storey above ground (m)[2]	Provide vehicle access to:	Type of appliance
Up to 2000	Up to 11 Over 11	See paragraph 15.1 15% of perimeter	Pump High reach
2000–8000	Up to 11 Over 11	15% of perimeter 50% of perimeter	Pump High reach
8000–16,000	Up to 11 Over 11	50% of perimeter 50% of perimeter	Pump High reach
16,000–24,000	Up to 11 Over 11	75% of perimeter 75% of perimeter	Pump High reach
Over 24,000	Up to 11 Over 11	100% of perimeter 100% of perimeter	Pump High reach

NOTES:

1. The sum of the area of all storeys in the building (excluding basements).
2. For storage buildings (purpose group 7(a)), measure height to mean roof level (see Appendix D).

See para 15.1

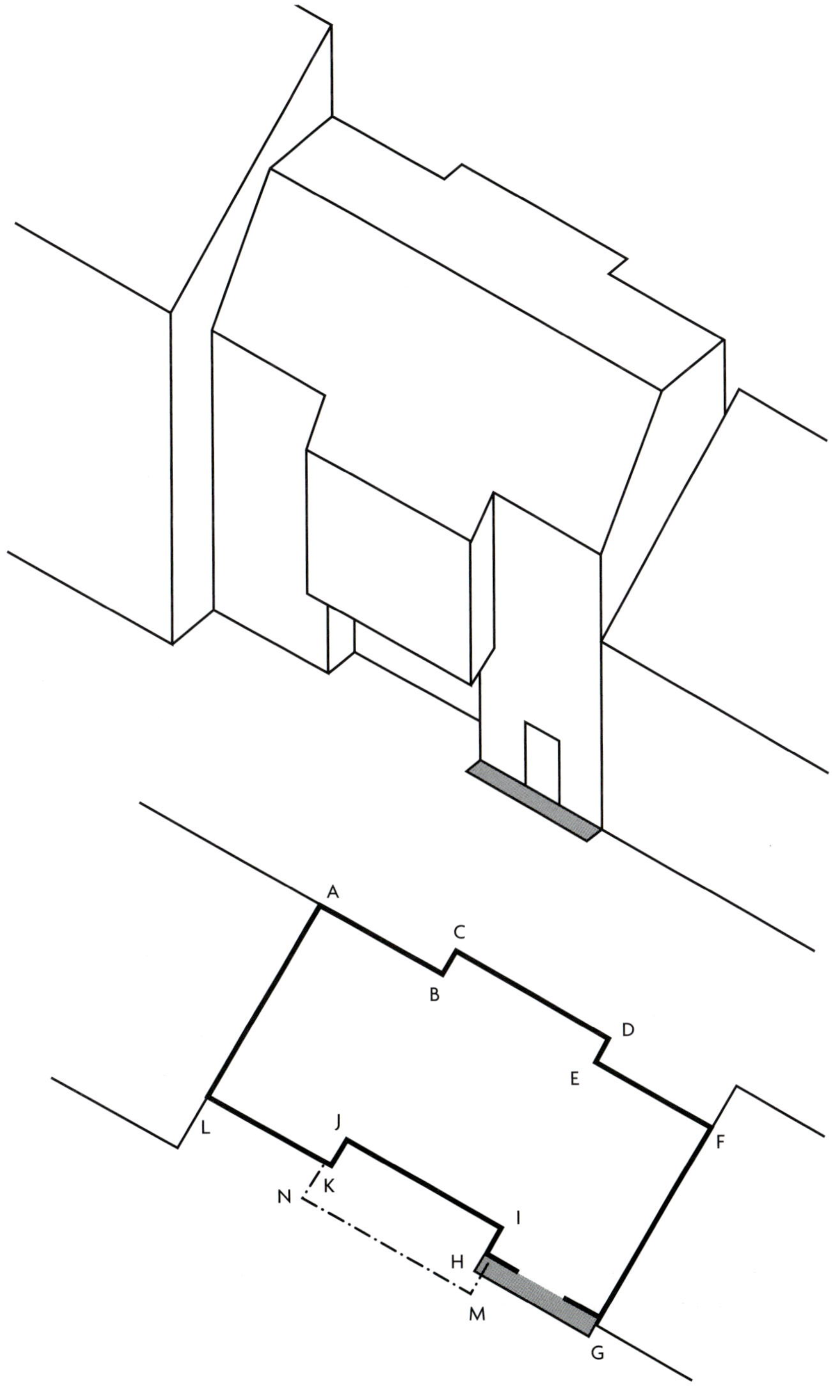

Plan of building AFGL where AL and FG are walls in common with other buildings.

The footprint of the building is the maximum aggregate plan perimeter found by the vertical projection of any overhanging storey onto a ground storey (i.e. ABCDEFGHMNKL).

The perimeter of the building for the purposes of Table 15.1 is the sum of the lengths of the two external walls, taking account of the footprint i.e. (A to B to C to D to E to F) + (G to H to M to N to K to L).

If the dimensions of the building are such that Table 15.1 requires vehicle access, the shaded area illustrates one possible example of 15% of the perimeter. Note: There should be a door into the building in this length (see paragraph 15.3).

If the building does not have walls in common with other buildings, the lengths AL and FG would be included in the perimeter.

Diagram 15.1 Example of building footprint and perimeter

Buildings fitted with fire mains

15.4 For buildings fitted with dry fire mains, both of the following apply.

a. Access should be provided for a pumping appliance to within 18m of each fire main inlet connection point. Inlets should be on the face of the building.

b. The fire main inlet connection point should be visible from the parking position of the appliance, and satisfy paragraph 16.10.

15.5 For buildings fitted with wet fire mains, access for a pumping appliance should comply with both of the following.

a. Within 18m, and within sight of, an entrance giving access to the fire main.

b. Within sight of the inlet to replenish the suction tank for the fire main in an emergency.

15.6 Where fire mains are provided in buildings for which Sections 16 and 17 make no provision, vehicle access may be as described in paragraphs 15.4 and 15.5, rather than Table 15.1.

Design of access routes and hardstandings

15.7 Access routes and hardstandings should comply with the guidance in Table 15.2. Requirements can only apply to the site of the works.

It may not be reasonable to upgrade the route across a site to a small building. The building control body, in consultation with the fire and rescue service, should consider options from doing no work to upgrading certain features, such as sharp bends.

15.8 Where access to an elevation is provided in accordance with Table 15.1, the following requirements should be met, depending on the building height.

a. Buildings up to 11m, excluding small buildings (paragraph 15.1): pump appliance access should be provided adjacent to the building for the specified percentage of the total perimeter.

b. Buildings over 11m: access routes should comply with the guidance in Diagram 15.2.

15.9 Where access is provided for high reach appliances in accordance with Table 15.1, overhead obstructions (such as cables and branches) should be avoided in the zone shown in Diagram 15.2.

15.10 Dead-end access routes longer than 20m require turning facilities, as in Diagram 15.3. Turning facilities should comply with the guidance in Table 15.2.

Table 15.2 Typical fire and rescue service vehicle access route specification

Appliance type	Minimum width of road between kerbs (m)	Minimum width of gateways (m)	Minimum turning circle between kerbs (m)	Minimum turning circle between walls (m)	Minimum clearance height (m)	Minimum carrying capacity (tonnes)
Pump	3.7	3.1	16.8	19.2	3.7	12.5
High reach	3.7	3.1	26.0	29.0	4.0	17.0

NOTES:

1. Fire appliances are not standardised. The building control body may, in consultation with the local fire and rescue service, use other dimensions.
2. The roadbase can be designed to 12.5 tonne capacity. Structures such as bridges should have the full 17-tonne capacity. The weight of high reach appliances is distributed over a number of axles, so infrequent use of a route designed to accommodate 12.5 tonnes should not cause damage.

See paras 15.8 and 15.9

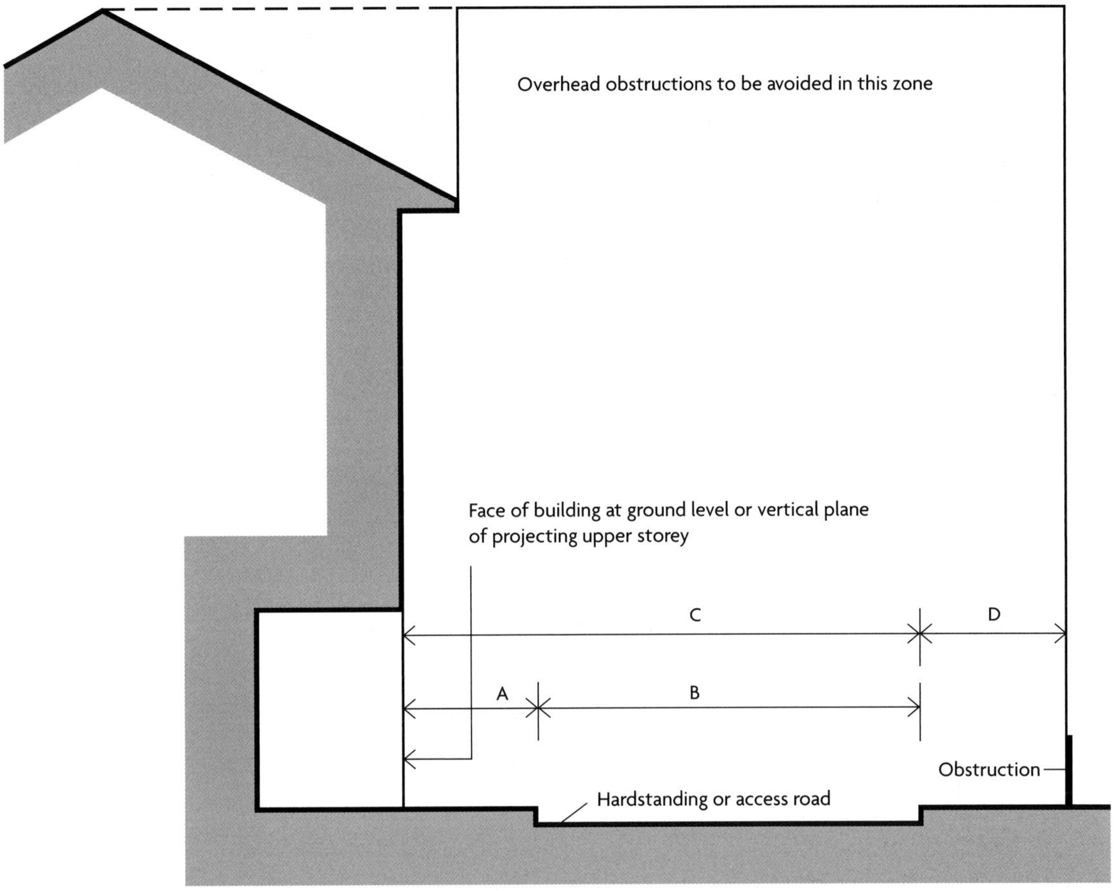

	Type of appliance	
	Turntable ladder dimension (m)	**Hydraulic platform dimension (m)**
A. Maximum distance of near edge of hardstanding from building	4.9	2.0
B. Minimum width of hardstanding	5.0	5.5
C. Minimum distance of further edge of hardstanding from building	10.0	7.5
D. Minimum width of unobstructed space (for swing of appliance platform)	N/A	2.2

NOTES:

1. Hardstanding for high reach appliances should be as level as possible and should have a maximum gradient of 1 in 12.
2. Fire appliances are not standardised. Some fire services have appliances with a greater weight or different size.
   In consultation with the fire and rescue service, the building control body should adopt the relevant dimensions and ground loading capacity.

Diagram 15.2 Relationship between building and hardstanding/access roads for high reach fire appliances

See para 15.10

Fire and rescue service vehicles should not have to reverse more than 20m from the end of an access road.

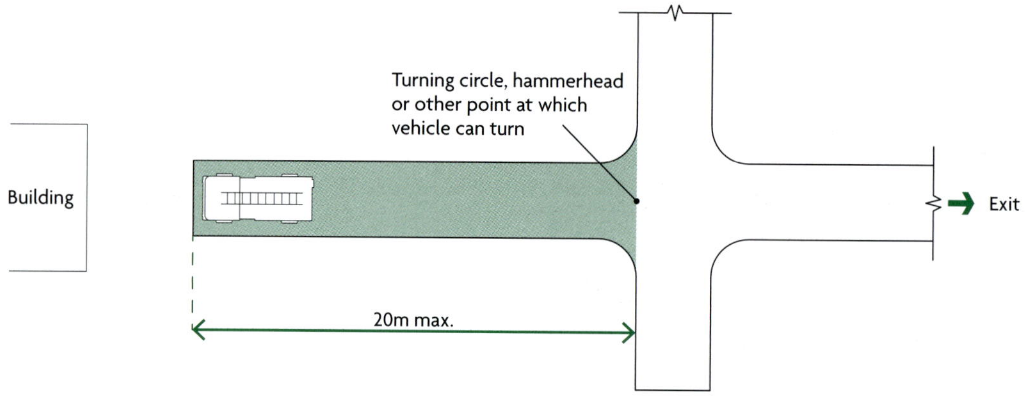

Diagram 15.3 Turning facilities

Section 16: Fire mains and hydrants

Introduction

16.1 Fire mains are installed for the fire and rescue service to connect hoses for water. They may be either of the following.

a. The 'dry' type, which are both of the following.

   i. Normally kept empty.

   ii. Supplied through a hose from a fire and rescue service pumping appliance.

b. The 'wet' type, which are both of the following.

   i. Kept full of water.

   ii. Supplied by pumps from tanks in the building.

There should be a facility to replenish a wet system from a pumping appliance in an emergency.

Provision of fire mains

16.2 Buildings with firefighting shafts should have fire mains in both of the following.

a. The firefighting shafts.

b. Where necessary, in protected escape stairs.

The criteria for providing firefighting shafts and fire mains are given in Section 17.

16.3 Buildings without firefighting shafts should be provided with fire mains where fire service vehicle access is not provided in accordance with Table 15.1. In these cases, outlets from fire mains should be located as described in paragraph 16.4, with a maximum hose distance of 45m from the fire main outlet to the furthest point, measured on a route suitable for laying a hose. Stairs do not need to be designed as firefighting shafts.

Design and construction of fire mains

16.4 If a firefighting shaft is provided, outlets from fire mains should be within the protected stairway or protected lobby (see Diagram 17.1).

16.5 Guidance on the design and construction of fire mains is given in **BS 9990**.

16.6 Buildings with a storey more than 50m above fire service vehicle access level should be provided with wet fire mains. In all other buildings where fire mains are provided, either wet or dry fire mains are suitable.

16.7 Fire service vehicle access to fire mains should be provided as described in paragraphs 15.4 and 15.5.

Provision of private hydrants

16.8 A building requires additional fire hydrants if both of the following apply.

a. It has a compartment with an area more than $280m^2$.

b. It is being erected more than 100m from an existing fire hydrant.

16.9 If additional hydrants are required, these should be provided in accordance with the following.

a. For buildings provided with fire mains – within 90m of dry fire main inlets.

b. For buildings not provided with fire mains – hydrants should be both of the following.

i. Within 90m of an entrance to the building.

ii. A maximum of 90m apart.

16.10 Each fire hydrant should be clearly indicated by a plate, fixed nearby in a conspicuous position, in accordance with **BS 3251**.

16.11 Guidance on aspects of provision and siting of private fire hydrants is given in **BS 9990**.

Alternative supply of water

16.12 An alternative source of water supply should be provided where any of the following apply.

a. No piped water supply is available.

b. Pressure and flow in the water main are insufficient.

c. An alternative source of supply is proposed.

16.13 The alternative source of water supply should be one of the following, subject to consultation with the local fire and rescue service.

a. A charged static water tank with a minimum capacity of 45,000 litres.

b. A spring, river, canal or pond that is capable of fulfilling both of the following conditions.

i. Providing or storing a minimum of 45,000 litres of water at all times.

ii. Providing access, space and a hardstanding for a pumping appliance.

c. Any other water supply that the local fire and rescue service considers appropriate.

Section 17: Access to buildings for firefighting personnel

Introduction

17.1 Facilities for fire and rescue, such as firefighting lifts, firefighting stairs and firefighting lobbies, are combined in protected firefighting shafts (Diagram 17.1). Section 8 gives guidance on the design and construction of protected shafts.

See para 17.1

Any building

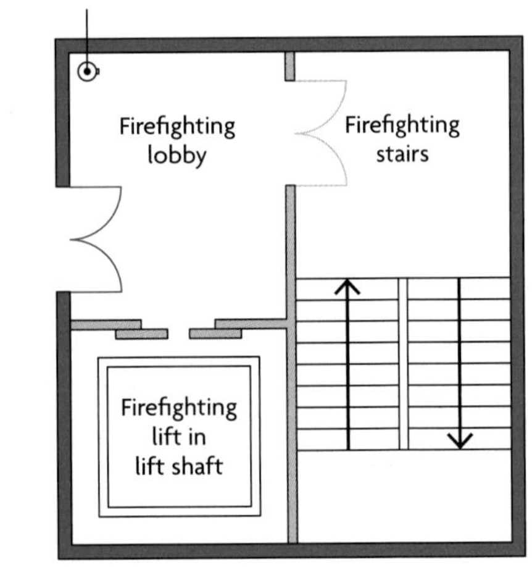

Minimum fire resistance REI 120 from accommodation side and REI 60 from inside the shaft with E 60 S_a fire doors

Minimum fire resistance REI 60 from both sides with E 30 S_a fire doors

NOTES:

1. Outlets from a fire main should be located in the firefighting lobby.
2. A firefighting lift is required if the building has a floor more than 18m above, or more than 10m below, fire service vehicle access level.
3. This diagram is only to illustrate the basic components and is not meant to represent the only acceptable layout. The firefighting shaft should be constructed generally in accordance with Section 6 of BS 9999.
4. For the minimum fire resistance of lift doors see Table C1.

Diagram 17.1 Components of a firefighting shaft

Provision of firefighting shafts

17.2 A building with a storey more than 18m above the fire and rescue service vehicle access level should have one or more firefighting shafts containing a firefighting lift. The number and location of firefighting shafts should comply with paragraphs 17.4 to 17.7. Firefighting shafts are not required to serve a basement that is not large or deep enough to need one (see paragraph 17.3 and Diagram 17.2).

17.3 A building with basement storeys should have firefighting shafts in accordance with the following.

a. There is a basement more than 10m below the fire and rescue service vehicle access level. The firefighting shafts should contain firefighting lifts.

b. There are two or more basement storeys, each with a minimum area of $900m^2$. The firefighting shafts do not need to include firefighting lifts.

The building's height and size determine whether firefighting shafts also serve upper storeys.

17.4 Firefighting shafts should serve all storeys through which they pass.

17.5 A minimum of two firefighting shafts should be provided to buildings with a storey that has both of the following.

a. A floor area of $900m^2$ or more.

b. A floor level 18m or more above the fire and rescue service vehicle access level.

17.6 At least two firefighting shafts, which do not need to include firefighting lifts, should be provided if buildings meet all of the following.

a. They are in the 'shop and commercial', 'assembly and recreation' or 'industrial' purpose group (purpose group 4, 5 or 6).

b. They have a storey area of $900m^2$ or more.

c. They have a storey height of 7.5m or more above fire and rescue service vehicle access level.

See para 17.2

Buildings in which firefighting shafts should be provided, showing which storeys need to be served

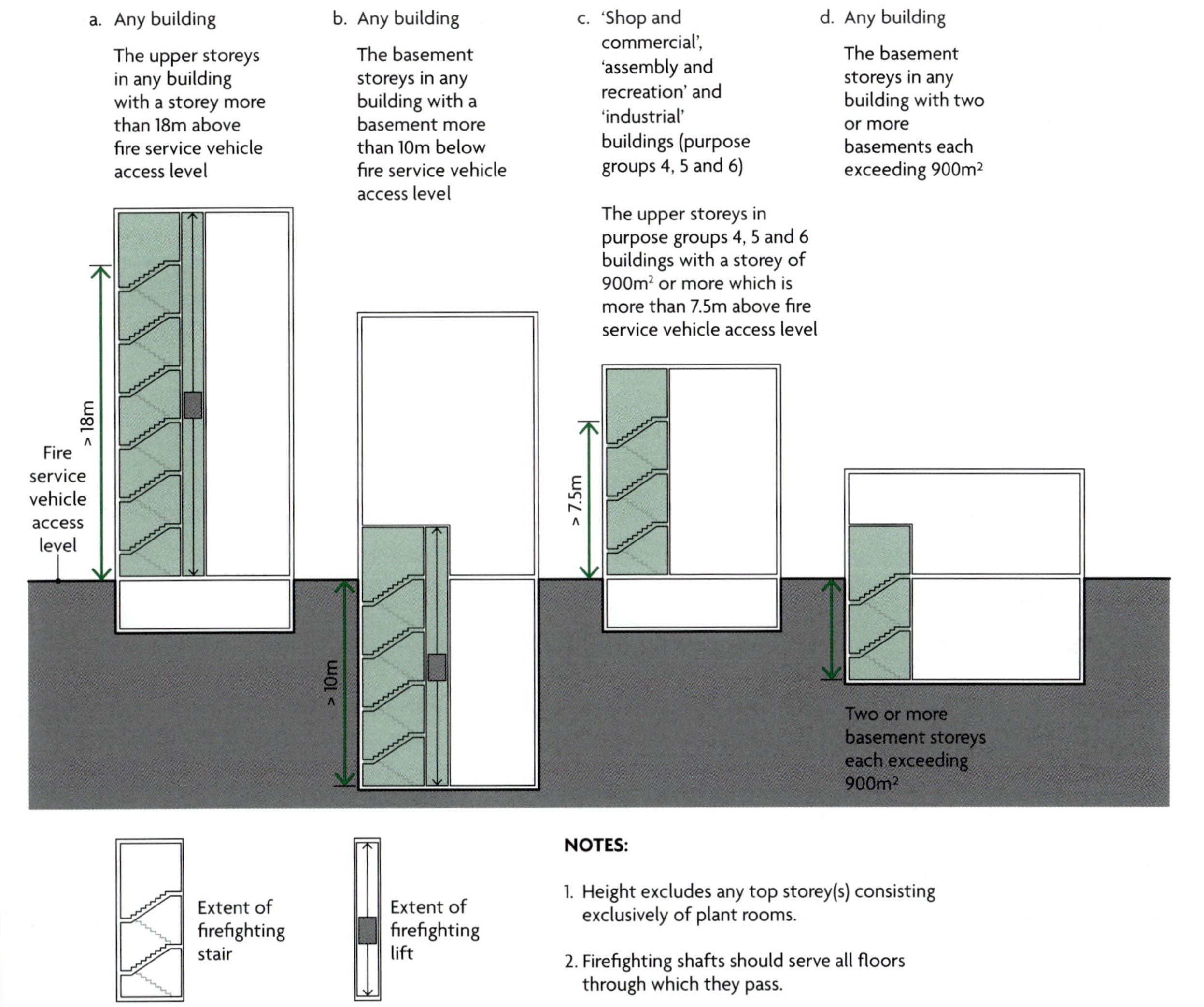

Diagram 17.2 Provision of firefighting shafts

Location of firefighting shafts

17.7 Firefighting shafts and protected stairways should be positioned such that every part of each storey more than 18m above the fire and rescue service vehicle access level complies with the maximum distances given in paragraph 17.8. Distances should be measured from the fire main outlet on a route suitable for laying a hose.

NOTE: If the internal layout is not known, the distance should be measured at two-thirds of the direct distance.

17.8 In any building, the hose laying distance should meet all of the following conditions.

a. A maximum of 60m from the fire main outlet in a firefighting shaft (see Diagram 17.3).

b. Additionally, where sprinklers have not been provided in accordance with Appendix E, the hose laying distance should be a maximum of 45m from a fire main outlet in a protected shaft (although this does not imply that the protected shaft needs to be designed as a firefighting shaft (see Diagram 17.3).

See para 17.8

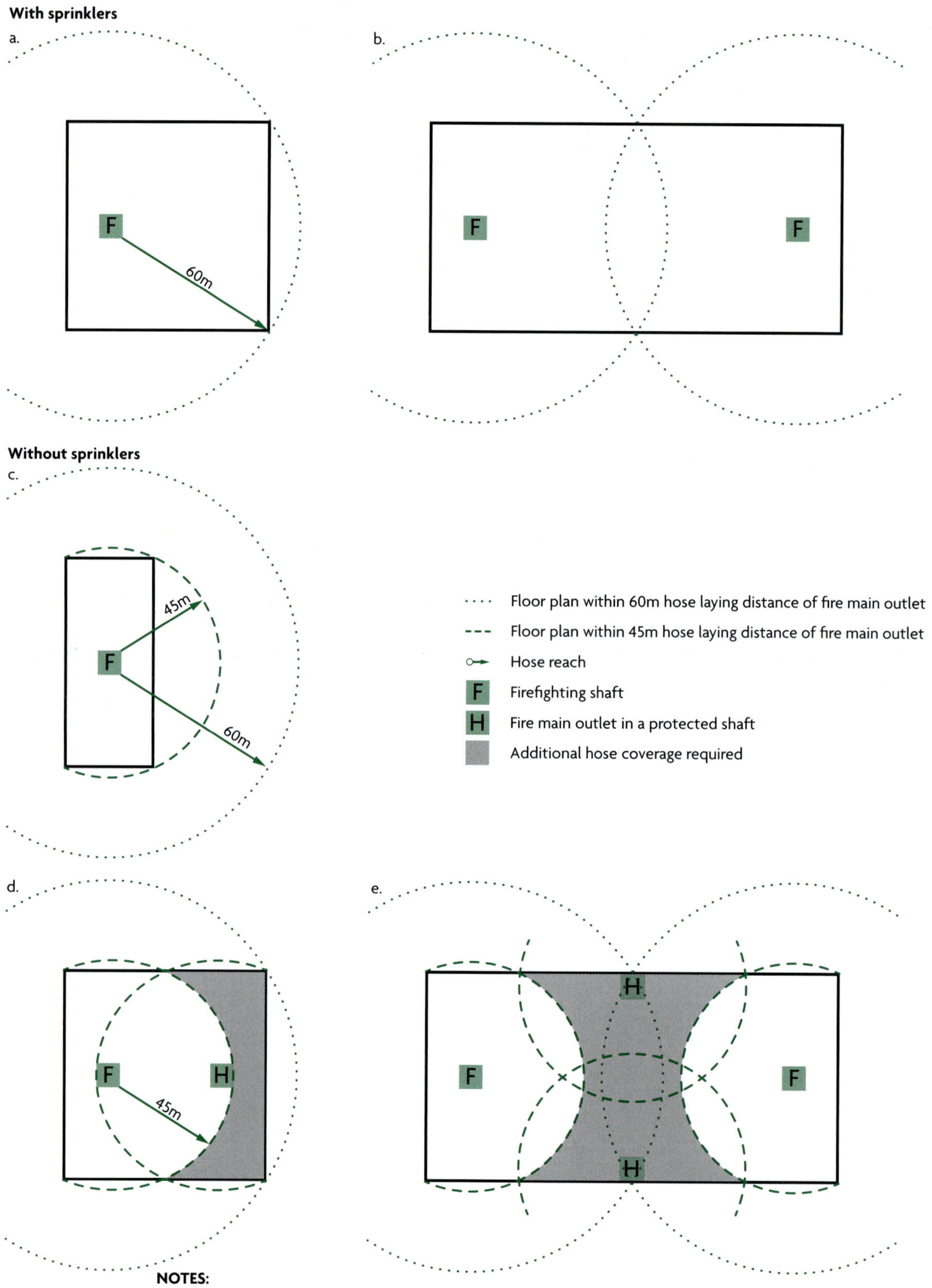

NOTES:

1. Hose laying distance should be measured from the fire main outlet along the route suitable for laying hose. If this route is not known, the distance should be taken at two-thirds of the direct distance.

2. The fire main outlet should be located according to paragraph 16.4.

Diagram 17.3 Location of firefighting shafts: hose laying distances

Design and construction of firefighting shafts

17.9 Every firefighting stair and firefighting lift should be approached from the accommodation through a firefighting lobby. Both the stair and lobby of the firefighting shaft should be provided with a means of venting smoke and heat (see clause 27.1 of **BS 9999**).

Only services associated with the firefighting shaft, such as ventilation systems and lighting for the firefighting shaft, should pass through or be contained within the firefighting shaft.

17.10 All firefighting shafts should have fire mains with outlet connections and valves at every storey.

17.11 A firefighting lift installation includes all of the following.

a. Lift car.

b. Lift well.

c. Lift machinery space.

d. Lift control system.

e. Lift communications system.

The lift shaft should be constructed in accordance with Section 6 of **BS 9999**.

Firefighting lift installations should conform to **BS EN 81-72** and **BS EN 81-20**.

Rolling shutters in compartment walls

17.12 The fire and rescue service should be able to manually open and close rolling shutters without the use of a ladder.

Section 18: Venting of heat and smoke from basements

Provision of smoke outlets

18.1 Heat and smoke from basement fires vented via stairs can inhibit access for firefighting personnel. This may be reduced by providing smoke outlets, or smoke vents, which allow heat and smoke to escape from the basement levels to the open air. They can also be used by the fire and rescue service to let cooler air into the basements (Diagram 18.1).

18.2 Each basement space should have one or more smoke outlets.

Where this is not practicable (for example, the plan area is deep and the amount of external wall is restricted by adjoining buildings), the perimeter basement spaces may be vented, with other spaces vented indirectly by opening connecting doors. This does not apply for places of special fire hazard (see paragraph 18.7).

If a basement is compartmented, each compartment should have one or more smoke outlets, rather than indirect venting.

A basement storey or compartment containing rooms with doors or windows does not need smoke outlets.

18.3 Smoke outlets connecting directly to the open air should be provided from every basement storey, except for any basement storey that has both of the following.

a. A maximum floor area of $200m^2$.

b. A floor a maximum of 3m below the adjacent ground level.

18.4 Strong rooms do not need to be provided with smoke outlets.

Natural smoke outlets

18.5 Smoke outlets should be both of the following.

a. Sited at high level in either the ceiling or wall of the space they serve.

b. Evenly distributed around the perimeter, to discharge to the open air.

18.6 The combined clear cross-sectional area of all smoke outlets should be a minimum of 1/40 of the area of the floor of the storey they serve.

18.7 Separate outlets should be provided from places of special fire hazard.

18.8 If the smoke outlet terminates at a point that is not readily accessible, it should be kept unobstructed and covered only with a class A1 grille or louvre.

18.9 If the smoke outlet terminates in a readily accessible position, it may be covered by a panel, stallboard or pavement light that can be broken out or opened. The position of covered smoke outlets should be suitably indicated.

18.10 Outlets should not be placed where they prevent the use of escape routes from the building.

Mechanical smoke extract

18.11 If basement storeys are fitted with a sprinkler system in accordance with Appendix E, a mechanical smoke extraction system may be provided as an alternative to natural venting. Sprinklers do not need to be installed on the other storeys unless needed for other reasons.

Car parks are not normally expected to be fitted with sprinklers (see Section 11 for guidance on car parks).

18.12 The air extraction system should comply with all of the following.

a. It should give at least 10 air changes per hour.

b. It should be capable of handling gas temperatures of 300°C for not less than one hour.

c. It should do either of the following.

   i. Be activated automatically if the sprinkler system activates.

   ii. Be activated by an automatic fire detection system that conforms to **BS 5839-1** (minimum L3 standard).

Further information on equipment for removing hot smoke is given in **BS EN 12101-3**.

See para 18.13

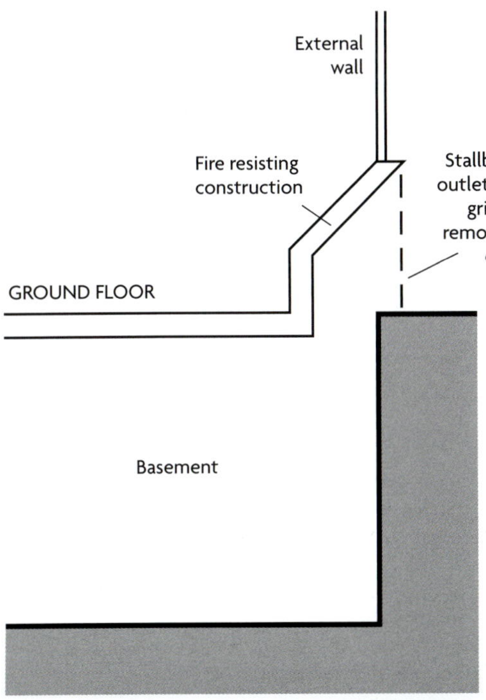

Diagram 18.1 Fire resisting construction for smoke outlet shafts

Construction of outlet ducts or shafts

18.13 Outlet ducts or shafts, including any bulkheads over them (see Diagram 18.1), should be enclosed in construction of class A1 rating and fire resistance at least equal to that of the element through which they pass.

18.14 Natural smoke outlet shafts should be separated from each other using construction of class A1 rating and fire resistance at least equal to that of the storeys they serve, where the shafts are either of the following.

a. From different compartments of the same basement storey.

b. From different basement storeys.

Basement car parks

18.15 The provisions for ventilation of basement car parks in Section 11 satisfy the requirements for venting smoke from any basement used as a car park.

Regulation 38: Fire safety information

This section deals with the following regulation of the Building Regulations 2010.

Fire safety information

38. (1) This regulation applies where building work—

(a) consists of or includes the erection or extension of a relevant building; or

(b) is carried out in connection with a relevant change of use of a building,

and Part B of Schedule 1 imposes a requirement in relation to the work.

(2) The person carrying out the work shall give fire safety information to the responsible person not later than the date of completion of the work, or the date of occupation of the building or extension, whichever is the earlier.

(3) In this regulation—

(a) "fire safety information" means information relating to the design and construction of the building or extension, and the services, fittings and equipment provided in or in connection with the building or extension which will assist the responsible person to operate and maintain the building or extension with reasonable safety;

(b) a "relevant building" is a building to which the Regulatory Reform (Fire Safety) Order 2005 applies, or will apply after the completion of building work;

(c) a "relevant change of use" is a material change of use where, after the change of use takes place, the Regulatory Reform (Fire Safety) Order 2005 will apply, or continue to apply, to the building; and

(d) "responsible person" has the meaning given by article 3 of the Regulatory Reform (Fire Safety) Order 2005.

Intention

The aim of this regulation is to ensure that the person responsible for the building has sufficient information relating to fire safety to enable them to manage the building effectively. The aim of regulation 38 will be achieved when the person responsible for the building has all the information to enable them to do all of the following.

a. Understand and implement the fire safety strategy of the building.

b. Maintain any fire safety system provided in the building.

c. Carry out an effective fire risk assessment of the building.

Section 19: Fire safety information

19.1 For building work involving the erection or extension of a relevant building (i.e. a building to which the Regulatory Reform (Fire Safety) Order 2005 applies or will apply), or the relevant change of use of a building, fire safety information should be given to the responsible person at one of the following times.

a. When the project is complete.

b. When the building or extension is first occupied.

19.2 This section is a guide to the information that should be provided. Guidance is in terms of essential information and additional information for complex buildings; however, the level of detail required should be considered on a case-by-case basis.

Essential information

19.3 Basic information on the location of fire protection measures may be sufficient. An as-built plan of the building should be provided showing all of the following.

a. Escape routes – this should include exit capacity (i.e. the maximum allowable number of people for each storey and for the building).

b. Location of fire-separating elements (including cavity barriers in walk-in spaces).

c. Fire doorsets, fire doorsets fitted with a self-closing device and other doors equipped with relevant hardware.

d. Locations of fire and/or smoke detector heads, alarm call points, detection/alarm control boxes, alarm sounders, fire safety signage, emergency lighting, fire extinguishers, dry or wet fire mains and other firefighting equipment, and hydrants outside the building.

e. Any sprinkler systems, including isolating valves and control equipment.

f. Any smoke control systems, or ventilation systems with a smoke control function, including mode of operation and control systems.

g. Any high risk areas (e.g. heating machinery).

19.4 Details should be provided of all of the following.

a. Specifications of any fire safety equipment provided, including routine maintenance schedules.

b. Any assumptions regarding the management of the building in the design of the fire safety arrangements.

c. Any provision enabling the evacuation of disabled people, which can be used when designing suitable personal emergency evacuation plans.

Additional information for complex buildings

19.5 A detailed record should be provided of both of the following.

a. The fire safety strategy.

b. Procedures for operating and maintaining any fire protection measures. This should include an outline cause and effect matrix/strategy for the building.

Further guidance is available in clause 9 and Annex H of **BS 9999**.

19.6 The records should include details of all of the following.

a. The fire safety strategy, including all assumptions in the design of the fire safety systems (such as fire load). Any risk assessments or risk analysis.

b. All assumptions in the design of the fire safety arrangements for the management of the building.

c. All of the following.

   i. Escape routes (including occupant load and capacity of escape routes).

   ii. Any provision to enable the evacuation of disabled people.

   iii. Escape strategy (e.g. simultaneous or phased).

   iv. Muster points.

d. All passive fire safety measures, including all of the following.

   i. Compartmentation (i.e. location of fire-separating elements).

   ii. Cavity barriers.

   iii. Fire doorsets, including fire doorsets fitted with a self-closing device and other doors equipped with relevant hardware (e.g. electronic security locks).

   iv. Duct dampers.

   v. Fire shutters.

e. All of the following.

   i. Fire detector heads.

   ii. Smoke detector heads.

   iii. Alarm call points.

   iv. Detection/alarm control boxes.

   v. Alarm sounders.

   vi. Emergency communications systems

   vii. CCTV.

   viii. Fire safety signage.

   ix. Emergency lighting.

   x. Fire extinguishers.

   xi. Dry or wet fire mains and other firefighting equipment.

   xii. Other interior facilities for the fire and rescue service.

    xiii. Emergency control rooms.
    xiv. Location of hydrants outside the building.
    xv. Other exterior facilities for the fire and rescue service.

f. All active fire safety measures, including both of the following.
    i. Sprinkler system(s) design, including isolating valves and control equipment.
    ii. Smoke control system(s) (or heating, ventilating and air conditioning system with a smoke control function) design, including mode of operation and control systems.

g. Any high-risk areas (e.g. heating machinery) and particular hazards.

h. Plans of the building as built, showing the locations of the above.

i. Both of the following.
    i. Specifications of any fire safety equipment provided, including all of the following.
        - Operational details.
        - Operators' manuals.
        - Software.
        - System zoning.
        - Routine inspection, testing and maintenance schedules.
    ii. Records of any acceptance or commissioning tests.

j. Any other details appropriate for the specific building.

Appendix A: Key terms

NOTE: Except for the items marked * (which are from the Building Regulations 2010), these definitions apply only to Approved Document B.

NOTE: The terms defined below are key terms used in this document only. Refer to **BS 4422** for further guidance on the definitions of common terms used in the fire safety industry which are not listed below.

Access room A room that the only escape route from an inner room passes through.

Alternative escape routes Escape routes that are sufficiently separated by direction and space or by fire resisting construction to ensure that one is still available if the other is affected by fire.

NOTE: A second stair, balcony or flat roof which enables a person to reach a place free from danger from fire is considered an alternative escape route for the purposes of a dwellinghouse.

Alternative exit One of two or more exits, each of which is separate from the other.

Appliance ventilation duct A duct to deliver combustion air to a gas appliance.

Atrium (plural **atria**) A continuous space that passes through one or more structural floors within a building, not necessarily vertically.

NOTE: Enclosed lift wells, enclosed escalator wells, building services ducts and stairs are not classified as atria.

Automatic release mechanism A device that normally holds a door open, but closes it automatically if any one of the following occurs.

- Smoke is detected by an automatic device of a suitable nature and quality in a suitable location.
- A hand-operated switch, fitted in a suitable position, is operated.
- The electricity supply to the device, apparatus or switch fails.
- The fire alarm system, if any, is operated.

Basement storey A storey with a floor that, at some point, is more than 1200mm below the highest level of ground beside the outside walls. (However, see Appendix B, paragraph B26c, for situations where the storey is considered to be a basement only because of a sloping site.)

Boundary The boundary of the land that belongs to a building, or, where the land abuts a road, railway, canal or river, the centre line of that road, railway, canal or river.

***Building** Any permanent or temporary building but not any other kind of structure or erection. A reference to a building includes a reference to part of a building.

Building control body A term that includes both local authority building control and approved inspectors.

Cavity A space enclosed by elements of a building (including a suspended ceiling) or contained within an element, but that *is not* a room, cupboard, circulation space, protected shaft, or space within a flue, chute, duct, pipe or conduit.

Cavity barrier A construction within a cavity, other than a smoke curtain, to perform either of the following functions.

- Close a cavity to stop smoke or flame entering.
- Restrict the movement of smoke or flame within a cavity.

Ceiling Part of a building that encloses a room, protected shaft or circulation space and is exposed overhead.

NOTE: The soffit of a rooflight, but not the frame, is included as part of the surface of the ceiling. An upstand below a rooflight is considered as a wall.

Circulation space A space (including a protected stairway) mainly used as a means of access between a room and an exit from the building or compartment.

Common balcony A walkway, open to the air on one or more sides, that forms part of the escape route from more than one flat.

Common stair An escape stair that serves more than one flat.

Compartment (fire) A building or part of a building, comprising one or more rooms, spaces or storeys, that is constructed to prevent the spread of fire to or from another part of the same building or an adjoining building.

NOTE: A roof space above the top storey of a compartment is included in that compartment. (See also 'Separated part'.)

Compartment wall or floor A fire resisting wall or floor to separate one fire compartment from another.

NOTE: Provisions relating to construction are given in Section 8.

Corridor access A design of a building containing flats, in which each flat is approached via a common horizontal internal access or circulation space, which may include a common entrance hall.

Dead end An area from which escape is possible in one direction only.

Direct distance The shortest distance from any point within the floor area to the nearest storey exit, measured within the external enclosures of the building, and ignoring walls, partitions and fittings other than the enclosing walls and partitions to protected stairways.

***Dwelling** Includes a dwellinghouse and a flat.

NOTE: A dwelling is a unit where one or more people live (whether or not as a sole or main residence) in either of the following situations.

- A single person or people living together as a family.
- A maximum of six people living together as a single household, including where care is provided for residents.

***Dwellinghouse** Does not include a flat or a building containing a flat.

Element of structure Any of the following.

- A member that forms part of the structural frame of a building, or any other beam or column.
- A loadbearing wall or loadbearing part of a wall.
- A floor.
- A gallery (but *not* a loading gallery, fly gallery, stage grid, lighting bridge, or any gallery provided for similar purposes or for maintenance and repair).
- An external wall.
- A compartment wall (including a wall that is common to two or more buildings).

NOTE: However, see the guidance to requirement B3, paragraph 7.3, for a list of structures that are *not* considered to be elements of structure.

Emergency lighting Lighting for use when the power supply to the normal lighting fails.

Escape lighting The part of the emergency lighting that is provided to ensure that the escape route is illuminated at all material times.

Escape route The route along which people can escape from any point in a building to a final exit.

Evacuation lift A lift that may be used to evacuate people in a fire.

Exit passageway A protected passageway that connects a protected stairway to a final exit.

NOTE: Exit passageways should be protected to the same standard as the stairway they serve.

***External wall** The external wall of a building includes all of the following.

- Anything located within any space forming part of the wall.
- Any decoration or other finish applied to any external (but not internal) surface forming part of the wall.
- Any windows and doors in the wall.

- Any part of a roof pitched at an angle of more than 70 degrees to the horizontal if that part of the roof adjoins a space within the building to which persons have access, but not access only for the purpose of carrying out repairs or maintenance.

Final exit The end of an escape route from a building that gives direct access to a street, passageway, walkway or open space, and is sited to ensure that people rapidly disperse away from the building so that they are no longer in danger from fire and/or smoke.

NOTE: Windows are not acceptable as final exits.

Fire alarm system Combination of components for giving an audible and/or other perceptible warning of fire.

Fire damper A mechanical or intumescent device within a duct or ventilation opening that operates automatically and is designed to resist the spread of fire.

Fire and smoke damper A fire damper which, in addition to the performance of the fire damper, resists the spread of smoke.

Fire doorset A door or shutter which, together with its frame and furniture as installed in a building, is intended (when closed) to resist the spread of fire and/or gaseous products of combustion and meets specified performance criteria to those ends.

NOTE: A fire doorset may have one or more leaves. The term includes a cover or other form of protection to an opening in a fire resisting wall or floor, or in a structure that surrounds a protected shaft. A fire doorset is a complete door assembly, assembled on site or delivered as a completed assembly, consisting of the door frame, leaf or leaves, essential hardware, edge seals and glazing, and any integral side panels or fanlight panels in an associated door screen.

Firefighting lift A lift with additional protection and with controls that enable it to be used by the fire and rescue service when fighting a fire. (See Section 17.)

Firefighting lobby A protected lobby that provides access from a firefighting stair to the accommodation area and to any associated firefighting lift.

Firefighting shaft A protected enclosure that contains a firefighting stair, firefighting lobbies and, if provided, a firefighting lift together with its machine room.

Firefighting stair A protected stairway that connects to the accommodation area through only a firefighting lobby.

Fire resisting (**Fire resistance**) The ability of a component or a building to satisfy, for a stated period of time, some or all of the appropriate criteria given in the relevant standard.

Fire-separating element A compartment wall, compartment floor, cavity barrier and construction that encloses a protected escape route and/or a place of special fire hazard.

Fire-stop (**Fire-stopping**) A seal provided to close an imperfection of fit or design tolerance between elements or components, to restrict the spread of fire and smoke.

***Flat** A flat is a separate and self-contained premises constructed or adapted for use for residential purposes and forming part of a building from some other part of which it is divided horizontally.

Gallery A floor or balcony that does not extend across the full extent of a building's footprint and is open to the floor below.

Habitable room A room used, or intended to be used, for people to live in (including, for the purposes of Approved Document B Volumes 1 and 2, a kitchen, but not a bathroom).

Height (of a building or storey for the purposes of Approved Document B Volumes 1 and 2)

- Height of a building is measured as shown in Appendix D, Diagram D4.
- Height of the floor of the top storey above ground level is measured as shown in Appendix D, Diagram D6.

Inner room Room from which escape is possible only by passing through another room (the access room).

Live/work unit A flat that is a workplace for people who live there, its occupants, and for people who do not live on the premises.

Means of escape Structural means that provide one or more safe routes for people to go, during a fire, from any point in the building to a place of safety.

Measurement

- Width of a doorway, cubic capacity, area, height of a building and number of storeys are measured as shown in Appendix D, Diagrams D1 to D6.
- Occupant number, travel distance, escape route and stairs are measured as described in Appendix D, paragraphs D1 to D4.

Notional boundary A boundary presumed to exist between two buildings on the same site.

Open spatial planning The internal arrangement of a building in which more than one storey or level is contained in one undivided volume, e.g. split-level floors. For the purposes of this document there is a distinction between open spatial planning and an atrium space.

Perimeter (of a building) The maximum aggregate plan perimeter, found by vertical projection onto a horizontal plane. (See Section 15.)

Pipe Includes pipe fittings and accessories. The definition of 'pipe' *excludes* a flue pipe and a pipe used for ventilating purposes, other than a ventilating pipe for an above-ground drainage system.

Place of special fire hazard A room such as any of the following.

- Oil-filled transformer room.
- Switch gear room.
- Boiler room.
- Storage space for fuel or other highly flammable substance(s).
- Room that houses a fixed internal combustion engine.

Platform floor (also called an access or raised floor) A floor that is supported by a structural floor, but with an intervening cavity to house services.

Protected circuit An electrical circuit that is protected against fire.

Protected corridor/lobby A corridor or lobby that is adequately protected from fire in adjoining areas by fire resisting construction.

Protected entrance hall/landing A circulation area, consisting of a hall or space in a flat, that is enclosed with fire resisting construction other than an external wall of a building.

Protected shaft A shaft that enables people, air or objects to pass from one compartment to another, and which is enclosed with fire resisting construction.

Protected stairway A stair that leads to a final exit to a place of safety and that is adequately enclosed with fire resisting construction. Included in the definition is any exit passageway between the foot of the stair and the final exit.

Purpose group A classification of a building according to the purpose to which it is intended to be put. (See Table 0.1.)

Relevant boundary The boundary or notional boundary that one side of the building faces and/or coincides with, and that is parallel or at an angle of a maximum of 80 degrees to that side of the building.

Rooflight A dome light, lantern light, skylight, ridge light, glazed barrel vault or other element to admit daylight through a roof.

Room An enclosed space within a building that is not used solely as a circulation space. The term includes not only conventional rooms, but also cupboards that are not fittings and large spaces such as warehouses and auditoria. The term *does not* include cavities such as ducts, ceiling cavities and roof spaces.

School A place of education for children between 2 and 19 years old. The term includes nursery schools, primary schools and secondary schools as defined in the Education Act 1996.

Self-closing device A device that closes a door, when open at any angle, against a door frame.

NOTE: If the door is in a cavity barrier, rising butt hinges (which are different from the self-closing device mentioned above) are acceptable.

Separated part (of a building) Part of a building that is separated from another part of the same building by a compartment wall. The wall runs the full height of the part and is in one vertical plane. (See Appendix D, Diagram D5.)

Sheltered housing Includes two or more dwellings in the same building or on adjacent sites, designed and constructed as residential accommodation for vulnerable or elderly people who receive, or will receive, a support service.

Single storey building A building that consists of a ground storey only. Basements are not counted as storeys in a building (see Appendix D). A separated part that consists of a ground storey only, with a roof to which access is only provided for repair or maintenance, may be treated as a single storey building.

Site (of a building) The land occupied by the building, up to the boundaries with land in other ownership.

***Specified attachment** Includes any of the following.

- A balcony attached to an external wall.
- A device for reducing heat gain within a building by deflecting sunlight which is attached to an external wall.
- A solar panel attached to an external wall.

Storey Includes any of the following.

- Any gallery in an assembly building (purpose group 5).
- Any gallery in any other type of building if its area is more than half that of the space into which it projects.
- A roof, unless it is accessible only for maintenance and repair.

NOTE: The building is regarded as a multi-storey building if both of the following apply.

- There is more than one gallery.
- The total aggregate area of all the galleries in one space is more than half the floor area of that space.

Storey exit A final exit, or a doorway that gives direct access into a protected stairway, firefighting lobby or external escape route.

NOTE: If an institutional building is planned to enable progressive horizontal evacuation, a door in a compartment wall is considered a storey exit for the purposes of requirement B1.

Suspended ceiling (fire-protecting) A ceiling suspended below a floor that adds to the fire resistance of the floor.

Thermoplastic material Any synthetic polymeric material that has a softening point below 200°C if tested to **BS EN ISO 306** Method A120. Specimens for this test may be fabricated from the original polymer where the thickness of material of the end product is less than 2.5mm.

Travel distance (unless otherwise specified, e.g. as in the case of flats) The distance that a person would travel from any point within the floor area to the nearest storey exit, determined by the layout of walls, partitions and fittings.

Unprotected area (in relation to a side or external wall of a building) All of the following are classed as unprotected areas.

- Any part of the external wall that has less than the relevant fire resistance set out in Section 13.
- Any part of the external wall constructed of material more than 1mm thick if that material does not have a class B-s3, d2 rating or better, which is attached or applied, whether for cladding or any other purpose.
- Windows, doors or other openings. This does not include windows that are designed and glazed to give the necessary level of fire resistance and that are not openable.

NOTE: Recessed car parking areas as shown in Diagram A1 should not be regarded as unprotected areas.

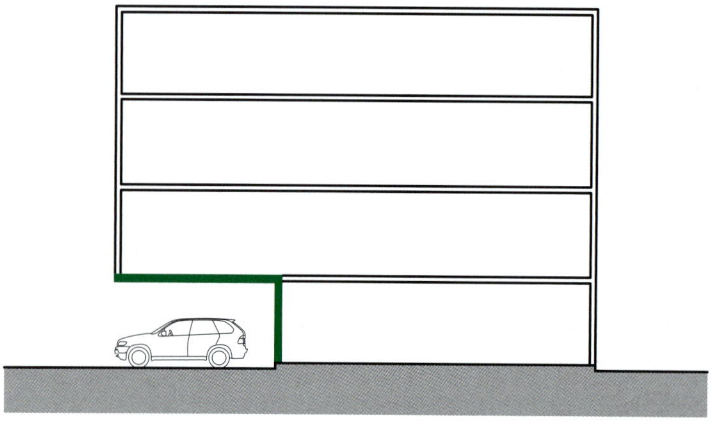

NOTE:

The parking area should be both of the following:

a. Open fronted.
b. Separated from the remainder of the building by a compartment wall(s) and floor(s) having not less than the period of fire resistance specified in Table B4 in Appendix B.

Diagram A1 Recessed car parking areas

Appendix B: Performance of materials, products and structures

Introduction

B1 Much of the guidance in this document is given in terms of performance classifications in relation to British or European Standards. In such cases, it will be necessary to demonstrate that a system or product can meet the relevant performance classification. This will be achieved if the system or product complies with one of the following.

a. They should be in accordance with a specification or design that has been shown by a specific test to be capable of meeting that performance classification.

b. They should have been designed by using relevant design standards in order to meet that performance classification.

c. They should have been assessed by applying relevant test evidence, in lieu of carrying out a specific test, as being capable of meeting that performance classification.

NOTE: Some products are subject to Classification Without Further Testing (CWFT). For the purposes of this approved document, such products can be considered to have been shown to be capable of meeting a performance specification as per paragraph B1a.

B2 Any test evidence used to demonstrate the fire performance classification of a product or system should be carefully checked to ensure that it is applicable to the intended use. Small differences in detail, such as fixing method, joints, dimensions, the introduction of insulation materials and air gaps (ventilated or not), can significantly affect the performance.

B3 Assessments should not be regarded as a way to avoid a test where one is necessary. Assessments should only be carried out where sufficient relevant test evidence is available. Relevant test evidence is unlikely to be provided by test standards which have different classification criteria.

B4 Where it is proposed to assess the classification of a product or system in lieu of carrying out a specific test (as in paragraph B1b), this should be done in accordance with the relevant standard for extended application for the test in question and should include details of the test evidence that has been used to support the assessment.

For performance classifications where there is no specific standard for extended application, assessment reports should be produced in accordance with the principles of **BS EN 15725** and should include details of the test evidence that has been used to support the assessment. Further information on best practice is provided in the Passive Fire Protection Federation's *Guide to Undertaking Assessments in Lieu of Fire Tests*.

NOTE: Regulation 7(2) limits components used in or on the external walls of certain buildings to materials achieving class A2-s1, d0 or class A1 (see Section 12). Assessments cannot be used to demonstrate compliance with this requirement.

B5 Tests and assessments should be carried out by organisations with the necessary expertise. For example, organisations listed as 'notified bodies' in accordance with the European Construction

Products Regulation or laboratories accredited by the United Kingdom Accreditation Service (UKAS) for the relevant test standard can be assumed to have the necessary expertise.

NOTE: Standard fire tests do not directly measure fire hazard. They measure or assess the response of a material or system to exposure to one or more aspects of fire conditions. Performance in fire tests is only one of a number of factors that should be taken into account.

Reaction to fire

B6 Reaction to fire relates to the degree to which a product will contribute, by its own decomposition, to a fire under specified conditions. Products, other than floorings, are classified as A1, A2, B, C, D, E or F (with class A1 being the highest performance and F being the lowest) in accordance with **BS EN 13501-1**. Class F is assigned when a product fails to attain class E. Untested products cannot be classified in accordance with **BS EN 13501-1**.

Materials covered by the Classification Without Further Testing (CWFT) process can be found by accessing the European Commission's website https://eur-lex.europa.eu/.

B7 The classes of reaction to fire performance of A2, B, C, D and E are accompanied by additional classifications related to the production of smoke (s1, s2, s3), with s1 indicating the lowest production, and/or flaming droplets/particles (d0, d1, d2), with d0 indicating the lowest production.

NOTE: When a classification includes s3, d2 this means that there is no limit set for smoke production and/or flaming droplets/particles.

B8 To reduce the testing burden on manufacturers, **BS EN 13238** defines a number of standard substrates that produce test results representative of different end use applications. The classification for reaction to fire achieved during testing is only valid when the product is used within this field of application, i.e. when the product is fixed to a substrate of that class in its end use. The standard substrate selected for testing should take account of the intended end use applications (field of application) of the product and represent end use substrates that have a density of a minimum of 75% of the standard substrate's nominal density.

B9 Standard substrates include gypsum plasterboard (**BS EN 520**) with a density of 700+/-100kg/m^3, calcium silicate board (**BS EN 14306**) 870+/-50kg/m^3 and fibre-cement board 1800+/-200kg/m^3.

NOTE: Standard calcium silicate board is not representative of gypsum plasterboard end use (due to the paper layer), but would be representative of most gypsum plasters (with densities of more than 650kg/m^3).

NOTE: Classifications based on tests using a plasterboard substrate would also be acceptable for products bonded to a gypsum plaster end use substrate.

National classifications for reaction to fire

B10 This document uses the European classification system for reaction to fire set out in **BS EN 13501-1**; however, there may be some products lawfully on the market using the classification system set out in previous editions. Where this is the case, Table B1 can be used for the purposes of this document.

Table B1 Reaction to fire classifications: transposition to national class

BS EN 13501-1 classification	Transposition
A1	Material that, when tested to **BS 476-11**, does not either: a. flame b. cause a rise in temperature on either the thermocouple at the centre of the specimen or in the furnaces
A2-s1, d0	None
A2-s3, d2	Material that meets either of the following. a. Any material of density 300kg/m^3 or more, which, when tested to **BS 476-11**, complies with both of the following: i. does not flame ii. causes a rise in temperature on the furnace thermocouple not exceeding 20°C b. Any material of density less than 300kg/m^3, which, when tested to **BS 476-11**, complies with both of the following: i. does not flame for more than 10 seconds ii. causes a rise in temperature on the thermocouple at the centre of the specimen or in the furnace that is a maximum of 35°C and on the furnace thermocouple that is a maximum of 25°C
B-s3, d2	Any material that meets both of the following criteria. a. Class 1 in accordance with **BS 476-7**. b. Has a fire propagation index (I) of a maximum of 12 and sub-index (i1) of a maximum of 6, determined by using the method given in **BS 476-6**. Index of performance (I) relates to the overall test performance, whereas sub-index (i1) is derived from the first three minutes of the test
C-s3, d2	Class 1 in accordance with **BS 476-7**
D-s3, d2	Class 3 in accordance with **BS 476-7**

NOTE: The national classifications do not automatically equate with the transposed classifications in the '**BS EN 13501-1** classification' column, therefore products cannot typically assume a European class unless they have been tested accordingly.

NOTE: A classification of s3, d2 indicates that no limit is set for production of smoke and/or flaming droplets/particles. If a performance for production of smoke and/or flaming droplets/particles is specified, then only the European classes can be used. For example, a national class may not be used as an alternative to a classification which includes s1, d0.

Thermoplastic materials

B11 Thermoplastic material is any synthetic polymeric material that has a softening point below 200°C if tested to **BS EN ISO 306** Method A120. Products formed from these materials cannot always be classified in the normal way. In those circumstances the following approach can be followed.

B12 Thermoplastic materials used for window glazing, rooflights and lighting diffusers within suspended ceilings do not need to meet the criteria within paragraph B19 onwards, if the guidance to requirements B2 and B4 is followed.

B13 For the purposes of requirements B2 and B4, thermoplastic materials should be classified as TP(a) rigid, TP(a) flexible or TP(b), as follows:

a. **TP(a) rigid**

i. rigid solid uPVC sheet

ii. solid (as distinct from double- or multi-skinned) polycarbonate sheet a minimum of 3mm thick

iii. multi-skinned rigid sheet made from uPVC or polycarbonate that has a class 1 rating when tested to **BS 476-7**

iv. any other rigid thermoplastic product, a specimen of which (at the thickness of the product as put on the market), when tested to **BS 2782-0** Method 508A, performs so that both:

- the test flame extinguishes before the first mark
- the duration of flaming or afterglow does not exceed 5 seconds following removal of the burner.

b. **TP(a) flexible**

Flexible products a maximum of 1mm thick that comply with the Type C requirements of **BS 5867-2** when tested to **BS 5438** Test 2 with the flame applied to the surface of the specimens for 5, 15, 20 and 30 seconds respectively, but excluding the cleansing procedure; and

c. **TP(b)**

i. rigid solid polycarbonate sheet products a maximum of 3mm thick, or multi-skinned polycarbonate sheet products that do not qualify as TP(a) by test

ii. other products which, when a specimen of the material between 1.5 and 3mm thick is tested in accordance with **BS 2782-0** Method 508A, have a maximum rate of burning of 50mm/minute.

NOTE: If it is not possible to cut or machine a 3mm thick specimen from the product, then a 3mm test specimen can be moulded from the same material as that used to manufacture the product.

B14 A thermoplastic material alone when used as a lining to a wall or ceiling cannot be assumed to protect a substrate. The surface rating of both thermoplastic material and substrate must therefore meet the required classification.

If, however, the thermoplastic material is fully bonded to a non-thermoplastic substrate, then only the surface rating of the composite needs to meet the required classification.

Roofs

B15 Performance of the resistance of roofs to external fire exposure is measured in terms of penetration through the roof construction and the spread of flame over its surface.

B16 Roof constructions are classified within the European system as B_{ROOF}(t4), C_{ROOF}(t4), D_{ROOF}(t4), E_{ROOF}(t4) or F_{ROOF}(t4) in accordance with **BS EN 13501-5**. B_{ROOF}(t4) indicates the highest performance and F_{ROOF}(t4) the lowest.

B17 **BS EN 13501-5** refers to four separate roof tests. The suffix (t4) used in paragraph B16 indicates that Test 4 is to be used for the purposes of this approved document.

B18 This document uses the European classification system for roof covering set out in **BS EN 13501-5**; however, there may be some products lawfully on the market using the classification system set out in previous editions. Where this is the case, Table B2 can be used for the purposes of this document.

Table B2 Roof covering classifications: transposition to national class

BS EN 13501-5 classification	Transposition to BS 476-3 classification
B_{ROOF} (t4)	AA, AB or AC
C_{ROOF}(t4)	BA, BB or BC
D_{ROOF}(t4)	CA, CB or CC
E_{ROOF}(t4)	AD, BD or CD
F_{ROOF}(t4)	DA, DB, DC or DD

NOTE: The national classifications do not automatically equate with the transposed classifications in the European column, therefore products cannot typically assume a European class unless they have been tested accordingly.

Fire resistance

B19 Common to all of the provisions of Part B of the Building Regulations is the property of fire resistance. Fire resistance is a measure of one or more of the following.

a. **Resistance to collapse** (loadbearing capacity), which applies to loadbearing elements only, denoted R in the European classification of the resistance to fire performance.

b. **Resistance to fire penetration** (integrity), denoted E in the European classification of the resistance to fire performance.

c. **Resistance to the transfer of excessive heat** (insulation), denoted I in the European classification of the resistance to fire performance.

B20 The standards of fire resistance necessary for a particular building are based on assumptions about the severity of fires and the consequences should an element fail. Fire severity is estimated in very broad terms from the use of the building (its purpose group), on the assumption that the building contents (which constitute the fire load) are similar for buildings with the same use.

B21 Because the use of buildings may change, a precise estimate of fire severity based on the fire load due to a particular use may be misleading. Therefore if a fire engineering approach of this kind is adopted, the likelihood that the fire load may change in the future needs to be considered.

B22 Performance in terms of the fire resistance to be achieved by elements of structure, doors and other forms of construction is classified in accordance with one of the following.

a. **BS EN 13501-2**.

b. **BS EN 13501-3**.

c. **BS EN 13501-4**.

B23 Fire resistance is measured in minutes. This relates to time elapsed in a standard test and should not be confused with real time.

B24 The fire resistance necessary for different circumstances is set out in the following tables.

a. Table B3 gives the specific requirements for each element of structure.

b. Table B4 sets out the minimum periods of fire resistance for elements of structure.

c. Table B5 sets out limitations on the use of uninsulated fire resisting glazed elements.

B25 This document uses the European classification system for fire resistance set out in **BS EN 13501-2** to **4**; however, there may be some products lawfully on the market using the classification system set out in previous editions. In those situations the equivalent classifications given in Table B1 can be used.

Table B3 Specific provisions of the test for fire resistance of elements of structure, etc.

Part of building	Minimum provisions when tested to the relevant European standard (minutes)[1]	Alternative minimum provisions when tested to the relevant part of **BS 476**[2] (minutes)			Type of exposure
		Loadbearing capacity[3]	Integrity	Insulation	
1. **Structural** frame, beam or column.	R see Table B4	See Table B4	Not applicable	Not applicable	Exposed faces
2. **Loadbearing wall** (which is not also a wall described in any of the following items).	R see Table B4	See Table B4	Not applicable	Not applicable	Each side separately
3. **Floors**[4]					
a. between a shop and flat above	REI 60 or see Table B4 (whichever is greater)	60 min or see Table B4 (whichever is greater)	60 min or see Table B4 (whichever is greater)	60 min or see Table B4 (whichever is greater)	From underside[5]
b. in upper storey of two storey dwellinghouse (but not over garage or basement)	R 30 and REI 15	30 min	15 min	15 min	From underside[5]
c. any other floor – including compartment floors.	REI see Table B4	See Table B4	See Table B4	See Table B4	From underside[5]

Table B3 Continued

Part of building	Minimum provisions when tested to the relevant European standard (minutes)[1]	Alternative minimum provisions when tested to the relevant part of **BS 476**[2] (minutes)			Type of exposure
		Loadbearing capacity[3]	Integrity	Insulation	
4. **Roofs**					
a. any part forming an escape route	REI 30	30 min	30 min	30 min	From underside[5]
b. any roof that performs the function of a floor.	REI see Table B4	See Table B4	See Table B4	See Table B4	From underside[5]
5. **External walls**					
a. any part a maximum of 1000mm from any point on the relevant boundary[6]	REI see Table B4	See Table B4	See Table B4	See Table B4	Each side separately
b. any part a minimum of 1000mm from the relevant boundary[6]	RE see Table B4 and REI 15	See Table B4	See Table B4	15 min	From inside the building
c. any part beside an external escape route (Section 2 Diagram 2.7 of Approved Document B Volume 1 and Section 3, Diagram 3.4).	RE 30	30 min	30 min	No provision[7] [8]	From inside the building
6. **Compartment walls** Separating either:					
a. a flat from any other part of the building (see paragraph 7.1 of Approved Document B Volume 1)	REI 60 or see Table B4 (whichever is less)	60 min or see Table B4 (whichever is less)	60 min or see Table B4 (whichever is less)	60 min or see Table B4 (whichever is less)	Each side separately
b. occupancies.	REI 60 or see Table B4 (whichever is less)	60 min or see Table B4 (whichever is less)	60 min or see Table B4 (whichever is less)	60 min or see Table B4 (whichever is less)	Each side separately
7. **Compartment walls** (other than in item 6 or item 10).	REI see Table B4	See Table B4	See Table B4	See Table B4	Each side separately

Table B3 Continued

Part of building	Minimum provisions when tested to the relevant European standard (minutes)[1]	Alternative minimum provisions when tested to the relevant part of **BS 476**[2] (minutes)			Type of exposure
		Loadbearing capacity[3]	Integrity	Insulation	
8. **Protected shafts** Excluding any firefighting shaft:					
a. any glazing	E 30	Not applicable	30 min	No provision[8]	Each side separately
b. any other part between the shaft and a protected lobby/corridor	REI 30	30 min	30 min	30 min	Each side separately
c. any part not described in (a) or (b) above.	REI see Table B4	See Table B4	See Table B4	See Table B4	Each side separately
9. **Enclosure** (that does not form part of a compartment wall or a protected shaft) to a:					
a. protected stairway	REI 30[8]	30 min	30 min	30 min[8]	Each side separately
b. lift shaft.	REI 30	30 min	30 min	30 min	Each side separately
10. **Wall or floor** separating an attached or integral garage from a dwellinghouse	REI 30[8]	30 min	30 min	30 min[8]	From garage side
11. **Fire resisting construction in dwellinghouses** not described elsewhere	REI 30[8]	30 min	30 min	30 min[8]	Each side separately
12. **Firefighting shafts**	REI 120	120 min	120 min	120 min	From side remote from shaft
a. construction that separates firefighting shaft from rest of building	REI 60	60 min	60 min	60 min	From shaft side
b. construction that separates firefighting stair, firefighting lift shaft and firefighting lobby.	REI 60	60 min	60 min	60 min	Each side separately

Table B3 Continued

Part of building	Minimum provisions when tested to the relevant European standard (minutes)[1]	Alternative minimum provisions when tested to the relevant part of **BS 476**[2] (minutes)			Type of exposure
		Loadbearing capacity[3]	Integrity	Insulation	
13. **Enclosure** (that is not a compartment wall or described in item 8) to a:					
a. protected lobby	REI 30[8]	30 min	30 min	30 min[8]	Each side separately
b. protected corridor.	REI 30[8]	30 min	30 min	30 min[8]	Each side separately
14. **Sub-division of a corridor**	REI 30[8]	30 min	30 min	30 min[8]	Each side separately
15. **Fire resisting construction**					
a. construction that encloses places of special fire hazard	REI 30	30 min	30 min	30 min	Each side separately
b. construction between store rooms and sales area in shops	REI 30	30 min	30 min	30 min	Each side separately
c. fire resisting sub-division	REI 30	30 min	30 min	30 min	Each side separately
d. construction that encloses bedrooms and ancillary accommodation in care homes.	REI 30	30 min	30 min	30 min	Each side separately
16. **Enclosure** in a flat to a protected entrance hall, or to a protected landing.	REI 30[8]	30 min	30 min	30 min[8]	Each side separately
17. **Cavity barrier**	E 30 and EI 15	Not applicable	30 min	15 min	Each side separately
18. **Ceiling** see paragraph 2.5, Diagram 2.3 of Approved Document B Volume 1 and paragraph 9.5 and Diagram 9.3.	EI 30	Not applicable	30 min	30 min	From underside
19. **Duct** described in paragraph 9.17e.	E 30	Not applicable	30 min	No provision	From outside
20. **Casing** around a drainage system described in Diagram 9.1 of Approved Document B Volume 1.	E 30	Not applicable	30 min	No provision	From outside

Table B3 Continued

Part of building	Minimum provisions when tested to the relevant European standard (minutes)[1]	Alternative minimum provisions when tested to the relevant part of **BS 476**[2] (minutes)			Type of exposure
		Loadbearing capacity[3]	Integrity	Insulation	
21. **Flue walls** described in Diagram 10.4.	EI half the period given in Table B4 for the compartment wall/floor	Not applicable	Half the period given in Table B4 for the compartment wall/floor	Half the period given in Table B4 for the compartment wall/floor	From outside
22. **Construction** described in note (a) to paragraph 12.9 of Approved Document B Volume 1.	EI 30	Not applicable	30 min	30 min	From underside
23. **Fire doorsets**	See Table C1	See Table C1			See Appendix C

NOTES:

1. **BS EN 13501-2** Classification using data from fire resistance tests, excluding ventilation services. **BS EN 13501-3** Classification using data from fire resistance tests on products and elements used in building service installations: fire resisting ducts and fire dampers. **BS EN 13501-4** Classification using data from fire resistance tests on components of smoke control systems.

   In the European classification:

   'R' is the resistance to fire in terms of loadbearing capacity.

   'E' is the resistance to fire in terms of integrity.

   'I' is the resistance to fire in terms of insulation.

   The national classifications do not automatically equate with the alternative classifications in the European column, therefore products cannot typically assume a European class unless they have been tested accordingly.
2. **BS 476-20** for general principles, **BS 476-21** for loadbearing elements, **BS 476-22** for non-loadbearing elements, **BS 476-23** for fire-protecting suspended ceilings and **BS 476-24** for ventilation ducts.
3. Applies to loadbearing elements only (see paragraph B19).
4. Guidance on increasing the fire resistance of existing timber floors is given in BRE Digest 208.
5. Only if a suspended ceiling meets the appropriate provisions should it be relied on to add to the fire resistance of the floor.
6. Such walls may contain areas that do not need to be fire resisting (unprotected areas). See Section 13.
7. Unless needed as part of a wall in item 5a or 5b.
8. Except for any limitations on uninsulated glazed elements given in Table B5.

Table B4 Minimum periods of fire resistance

Purpose group of building	Minimum periods of fire resistance[1] (minutes) in a:					
	Basement storey* including floor over		Ground or upper storey			
	Depth (m) of the lowest basement		Height (m) of top floor above ground, in a building or separated part of a building			
	More than 10	Up to 10	Up to 5	Up to 18	Up to 30	More than 30
1. Residential:						
a. Block of flats						
– without sprinkler system	90 min	60 min	30 min†	60 min+§	90 min+	Not permitted[2]
– with sprinkler system[3]	90 min	60 min	30 min†	60 min+§	90 min+	120 min+
b. and c. Dwellinghouse	Not applicable[4]	30 min*†	30 min†	60 min[5]	Not applicable[4]	Not applicable[4]
2. Residential						
a. Institutional	90 min	60 min	30 min†	60 min	90 min	120 min‡
b. Other residential	90 min	60 min	30 min†	60 min	90 min	120 min‡
3. Office:						
– without sprinkler system	90 min	60 min	30 min†	60 min	90 min	Not permitted[6]
– with sprinkler system[3]	60 min	60 min	30 min†	30 min†	60 min	120 min‡
4. Shop and commercial:						
– without sprinkler system	90 min	60 min	60 min	60 min	90 min	Not permitted
– with sprinkler system[3]	60 min	60 min	30 min†	60 min	60 min	120 min‡
5. Assembly and recreation:						
– without sprinkler system	90 min	60 min	60 min	60 min	90 min	Not permitted
– with sprinkler system[3]	60 min	60 min	30 min†	60 min	60 min	120 min‡
6. Industrial:						
– without sprinkler system	120 min	90 min	60 min	90 min	120 min	Not permitted
– with sprinkler system[3]	90 min	60 min	30 min†	60 min	90 min	120 min‡
7. Storage and other non-residential:						
a. any building or part not described elsewhere:						
– without sprinkler system	120 min	90 min	60 min	90 min	120 min	Not permitted
– with sprinkler system[3]	90 min	60 min	30 min†	60 min	90 min	120 min‡

Table B4 Continued

Purpose group of building	Minimum periods of fire resistance[1] (minutes) in a:					
	Basement storey* including floor over		Ground or upper storey			
	Depth (m) of the lowest basement		Height (m) of top floor above ground, in a building or separated part of a building			
	More than 10	Up to 10	Up to 5	Up to 18	Up to 30	More than 30
b. car park for light vehicles:						
i. open sided car park[7]	Not applicable	Not applicable	15 min†#	15 min†#(8)	15 min†#(8)	60 min
ii. any other car park	90 min	60 min	30 min†	60 min	90 min	120 min‡

NOTES:

For single storey buildings, the periods under the heading 'Up to 5' apply. If single storey buildings have basements, for the basement storeys the period appropriate to their depth applies.

* For the floor over a basement or, if there is more than one basement, the floor over the topmost basement, the higher of the period for the basement storey and the period for the ground or upper storey applies.

† For compartment walls that separate buildings, the period is increased to a minimum of 60 minutes.

\+ For any floor that does not contribute to the support of the building within a flat of more than one storey, the period is reduced to 30 minutes.

§ For flat conversions, refer to paragraphs 6.5 to 6.7 of Approved Document B Volume 1 regarding the acceptability of 30 minutes.

‡ For elements that do not form part of the structural frame, the period is reduced to 90 minutes.

\# For elements that protect the means of escape, the period is increased to 30 minutes.

1. Refer to note 1, Table B3 for the specific provisions of test.
2. Blocks of flats with a floor more than 30m above ground level should be fitted with a sprinkler system in accordance with Appendix E.

   NOTE: Sprinklers only need to be provided within the individual flats, they are not required in the common areas such as stairs, corridors or landings when these areas are fire sterile.
3. 'With sprinkler system' means that the building is fitted throughout with an automatic sprinkler system in accordance with Appendix E.
4. Very large (over 18m in height or with a 10m deep basement) or unusual dwellinghouses are outside the scope of the guidance provided with regard to dwellinghouses.
5. A minimum of 30 minutes in the case of three storey dwellinghouses, increased to 60 minutes minimum for compartment walls separating buildings.
6. Buildings within the 'office', 'shop and commercial', 'assembly and recreation', 'industrial' and 'storage and other non-residential' (except car parks for light vehicles) purpose groups (purpose groups 3 to 7(a)) require sprinklers where there is a top storey above 30m. The sprinkler system should be provided in accordance with Appendix E.
7. The car park should comply with the relevant provisions in the guidance on requirement B3, Section 11.
8. For the purposes of meeting the Building Regulations, the following types of steel elements are deemed to have satisfied the minimum period of fire resistance of 15 minutes when tested to the European test method.

   i. Beams supporting concrete floors, maximum $Hp/A=230m^{-1}$ operating under full design load.

   ii. Free-standing columns, maximum $Hp/A=180m^{-1}$ operating under full design load.

   iii. Wind bracing and struts, maximum $Hp/A=210m^{-1}$ operating under full design load.

   Guidance is also available in **BS EN 1993-1-2**.

Application of the fire resistance standards in Table B4

B26 The following guidance should be used when applying the fire resistance standards in Table B4.

a. If one element of structure supports or carries or gives stability to another, the fire resistance of the supporting element should be no less than the minimum period of fire resistance for the other element (whether that other element is loadbearing or not). In some circumstances, it may be reasonable to vary this principle, for example:

   i. if the supporting structure is in the open air and is not likely to be affected by the fire in the building

   ii. if the supporting structure is in a different compartment, with a fire-separating element (that has the higher standard of fire resistance) between the supporting and the separated structure

   iii. if a plant room on the roof needs greater fire resistance than the elements of structure that support it.

b. If an element of structure forms part of more than one building or compartment, that element should be constructed to the standard of the higher of the relevant provisions.

c. If, due to the slope of the ground, one side of a basement is open at ground level (allowing smoke to vent and providing access for firefighting) for elements of structure in that storey it may be appropriate to adopt the standard of fire resistance that applies to above-ground structures.

d. Although most elements of structure in a single storey building may not need fire resistance, fire resistance is needed if one of the following applies to the element.

   i. It is part of, or supports, an external wall, and there is provision in the guidance on requirement B4 to limit the extent of openings and other unprotected areas in the wall.

   ii. It is part of, or supports, a compartment wall, including a wall that is common to two or more buildings.

   iii. It supports a gallery.

B27 For the purposes of this paragraph, the ground storey of a building that has one or more basement storeys and no upper storeys may be considered as a single storey building. The fire resistance of the basement storeys should be that specified for basements.

Table B5 Limitations on the use of uninsulated glazed elements on escape routes. These limitations *do not* apply to glazed elements that satisfy the relevant insulation criterion, see Table B3

Position of glazed element	Maximum total glazed area in parts of a building with access to:			
	A single stair		More than one stair	
	Walls	Door leaf	Walls	Door leaf
Flats				
1. Within the enclosures of a protected entrance hall or protected landing, or within fire resisting separation shown in Section 3, Diagram 3.4, of Approved Document B Volume 1.	Fixed fanlights only	Unlimited above 1100mm from floor	Fixed fanlights only	Unlimited above 1100mm from floor
Dwellinghouses				
2. Within either: a. the enclosures of a protected stairway b. fire resisting separation shown in Diagram 2.2 of Approved Document B Volume 1.	Unlimited above 1100mm from floor or pitch of the stair	Unlimited	Unlimited above 1100mm from floor or pitch of the stair	Unlimited
3. Within fire resisting separation either: a. shown in Diagram 2.4 of Approved Document B Volume 1 b. described in paragraph 2.16b of Approved Document B Volume 1.	Unlimited above 100mm from floor	Unlimited above 100mm from floor	Unlimited above 100mm from floor	Unlimited above 100mm from floor
4. Existing window between an attached/integral garage and the dwellinghouse.	Unlimited	Not applicable	Unlimited	Not applicable
5. Adjacent to an external escape stair (see paragraph 2.17 and Diagram 2.7 of Approved Document B Volume 1) or roof escape route (see paragraph 2.13 of Approved Document B Volume 1).	Unlimited	Unlimited	Unlimited	Unlimited
General (except dwellinghouses)				
6. Between residential/sleeping accommodation and a common escape route (corridor, lobby or stair).	Nil	Nil	Nil	Nil
7. Between a protected stairway[1] and either: a. the accommodation b. a corridor that *is not* a protected corridor *other than in item 6 above.*	Nil	25% of door area	Unlimited above 1100mm[2]	50% of door area
8. Between either: a. a protected stairway[1] and a protected lobby or protected corridor b. accommodation and a protected lobby *other than in item 6 above.*	Unlimited above 1100mm from floor	Unlimited above 100mm from floor	Unlimited above 100mm from floor	Unlimited above 100mm from floor

Table B5 Continued

Position of glazed element	Maximum total glazed area in parts of a building with access to:			
	A single stair		More than one stair	
	Walls	Door leaf	Walls	Door leaf
9. Between the accommodation and a protected corridor that forms a dead end, *other than in item 6 above.*	Unlimited above 1100mm from floor	Unlimited above 100mm from floor	Unlimited above 1100mm from floor	Unlimited above 100mm from floor
10. Between accommodation and any other corridor, or subdividing corridors, *other than in item 6 above.*	Not applicable	Not applicable	Unlimited above 100mm from floor	Unlimited above 100mm from floor
11. Beside an external escape route.	Unlimited above 1100mm from floor	Unlimited above 1100mm from floor	Unlimited above 1100mm from floor	Unlimited above 1100mm from floor
12. Beside an external escape stair (see paragraph 3.32 and Diagram 3.4) or roof escape route (see paragraph 2.32).	Unlimited	Unlimited	Unlimited	Unlimited

NOTES:

Items 1 and 8 apply also to single storey buildings.

Fire resisting glass should be marked with the name of the manufacturer and the name of the product.

Further guidance can be found in *A Guide to Best Practice in the Specification and Use of Fire-resistant Glazed Systems* published by the Glass and Glazing Federation.

1. If the protected stairway is also a protected shaft or a firefighting stair (see Section 17), there may be further restrictions on the use of glazed elements.
2. Measured vertically from the landing floor level or the stair pitch line.
3. The 100mm limit is intended to reduce the risk of fire spreading from a floor covering.

C

Appendix C: Fire doorsets

C1 All fire doorsets should have the performance shown in Table C1, based on one of the following.

a. Fire resistance in terms of integrity, for a period of minutes, when tested to **BS 476-22**, e.g. FD 30. A suffix (S) is added for doorsets where restricted smoke leakage at ambient temperatures is needed.

b. As determined with reference to Commission Decision 2000/367/EC regarding the classification of the resistance to fire performance of construction products, construction works and parts thereof. All fire doorsets should be classified in accordance with **BS EN 13501-2**, tested to the relevant European method from the following.

   i. **BS EN 1634-1**.

   ii. **BS EN 1634-2**.

   iii. **BS EN 1634-3**.

c. As determined with reference to European Parliament and Council Directive 95/16/EC (which applies to lifts that permanently serve buildings and constructions and specified safety components) on the approximation of laws of Member States relating to lifts ('Lifts Directive') implementing the Lifts Regulations 1997 (SI 1997/831) and calling upon the harmonised standard **BS EN 81-58**.

C2 The performance requirement is in terms of integrity (E) for a period of minutes. An additional classification of S_a is used for all doors where restricted smoke leakage at ambient temperatures is needed.

C3 The requirement is for test exposure from each side of the doorset separately. The exception is lift doors, which are tested from the landing side only.

C4 Any test evidence used to verify the fire resistance rating of a doorset or shutter should be checked to ensure both of the following.

a. It adequately demonstrates compliance.

b. It is applicable to the **complete installed assembly**. Small differences in detail may significantly affect the rating.

Until relevant harmonised product standards are published, for the purposes of meeting the Building Regulations, products tested in accordance with **BS EN 1634-1** (with or without pre-fire test mechanical conditioning) that achieve the minimum performance in Table C1 will be deemed to satisfy the provisions.

C5 All fire doorsets, including to flat entrances and between a dwellinghouse and an integral garage, should be fitted with a self-closing device, except for all of the following.

a. Fire doorsets to cupboards.

b. Fire doorsets to service ducts normally locked shut.

c. Fire doorsets within flats and dwellinghouses.

C6 If a self-closing device would be considered to interfere with the normal approved use of the building, self-closing fire doors may be held open by one of the following.

a. A fusible link, but not if the doorset is in an opening provided as a means of escape unless it complies with paragraph C7.

b. An automatic release mechanism activated by an automatic fire detection and alarm system.

c. A door closer delay device.

C7 Two fire doorsets may be fitted in the same opening if each door is capable of closing the opening, so the total fire resistance is the sum of their individual resistances. If the opening is provided as a means of escape, both fire doorsets should be self-closing.

If one fire doorset is capable of being easily opened by hand and has a minimum of 30 minutes' fire resistance, the other fire doorset should comply with both of the following.

a. Be fitted with an automatic self-closing device.

b. Be held open by a fusible link.

C8 Fire doorsets often do not provide any significant insulation. Unless providing both integrity and insulation in accordance with Appendix B, Table B3, a maximum of 25% of the length of a compartment wall should consist of door openings.

Where it is practicable to maintain a clear space on both sides of the doorway, the above percentage may be greater.

C9 Rolling shutters should be capable of manual opening and closing for firefighting purposes (see Section 17). Rolling shutters across a means of escape should only be released by a heat sensor, such as a fusible link or electric heat detector, in the immediate vicinity of the door.

Unless a shutter is also intended to partially descend as part of a boundary to a smoke reservoir, shutters across a means of escape should not be closed by smoke detectors or a fire alarm system.

C10 Unless shown to be satisfactory when tested as part of a fire doorset assembly, the essential components of any hinge on which a fire door is hung should be made entirely from materials that have a minimum melting point of 800°C.

C11 Except for doorsets listed in paragraph C12, all fire doorsets should be marked with one of the following fire safety signs, complying with **BS 5499-5**, as appropriate.

a. To be kept closed when not in use – mark 'Fire door keep shut'.

b. To be kept locked when not in use – mark 'Fire door keep locked shut'.

c. Held open by an automatic release mechanism or free swing device – mark 'Automatic fire door keep clear'.

All fire doorsets should be marked on both sides, except fire doorsets to cupboards and service ducts, which should be marked on the outside.

C12 The following fire doorsets are not required to comply with paragraph C11.

a. Doors to and within flats and dwellinghouses.

b. Bedroom doors in 'residential (other)' (purpose group 2(b)) premises.

c. Lift entrance/landing doors.

C13 The performance of some doorsets set out in Table C1 is linked to the minimum periods of fire resistance for elements of structure given in Tables B3 and B4. Limitations on the use of uninsulated glazing in fire doorsets are given in Table B5.

C14 Recommendations for the specification, design, construction, installation and maintenance of fire doorsets constructed with non-metallic door leaves are given in **BS 8214**.

Guidance on timber fire resisting doorsets, in relation to the new European test standard, may be found in *Timber Fire Resisting Doorsets: Maintaining Performance Under the New European Test Standard* published by the Timber Research and Development Association (TRADA).

Guidance for metal doors is given in *Code of Practice for Fire Resisting Metal Doorsets* published by the Door and Shutter Manufacturers' Association (DSMA).

C15 Hardware used on fire doors can significantly affect their performance in a fire. Notwithstanding the guidance in this approved document, guidance is available in *Hardware for Fire and Escape Doors* published by the Door and Hardware Federation (DHF) and Guild of Architectural Ironmongers (GAI).

Table C1 Provisions for fire doorsets

Position of door	Minimum fire resistance of door in terms of integrity (minutes) when tested to the relevant European standard[1]	Minimum fire resistance of door in terms of integrity (minutes) when tested to **BS 476-22**
1. **In a compartment wall separating buildings**	Same as for the wall in which the door is fitted, but a minimum of 60 minutes	Same as for the wall in which the door is fitted, but a minimum of 60 minutes
2. **In a compartment wall:**		
a. if it separates a flat from a space in common use	E 30 S_a[2]	FD 30 S[2]
b. enclosing a protected shaft forming a stairway wholly or partly above the adjoining ground in a building used for flats, other residential, assembly and recreation, or office purposes	E 30 S_a[2]	FD 30 S[2]
c. enclosing a protected shaft forming a stairway not described in (b) above	Half the period of fire resistance of the wall in which it is fitted, but 30 minutes minimum and with suffix S_a[2]	Half the period of fire resistance of the wall in which it is fitted, but 30 minutes minimum and with suffix S[2]
d. enclosing a protected shaft forming a lift or service shaft	Half the period of fire resistance of the wall in which it is fitted, but 30 minutes minimum	Half the period of fire resistance of the wall in which it is fitted, but 30 minutes minimum
e. not described in (a), (b), (c) or (d) above.	Same as for the wall in which it is fitted, but add S_a[2] if the door is used for progressive horizontal evacuation under the guidance to requirement B1	Same as for the wall in which it is fitted, but add S[2] if the door is used for progressive horizontal evacuation under the guidance to requirement B1
3. **In a compartment floor**	Same as for the floor in which it is fitted	Same as for the floor in which it is fitted

Table C1 Continued

Position of door	Minimum fire resistance of door in terms of integrity (minutes) when tested to the relevant European standard[1]	Minimum fire resistance of door in terms of integrity (minutes) when tested to **BS 476-22**
4. **Forming part of the enclosures of:**		
a. a protected stairway (except as described in item 9 or 11(b) below)	E 30 S_a[2]	FD 30 S[2]
b. a lift shaft (see paragraph 5.34b) that does not form a protected shaft in 2(b), (c) or (d) above.	E 30	FD 30
5. **Forming part of the enclosure of:**		
a. a protected lobby approach (or protected corridor) to a stairway	E 30 S_a[2]	FD 30 S[2]
b. any other protected corridor	E 20 S_a[2]	FD 20 S[2]
c. a protected lobby approach to a lift shaft (see paragraph 5.37 to 5.39).	E 30 S_a[2]	FD 30 S[2]
6. **Giving access to an external escape route**	E 30	FD 30
7. **Sub-dividing:**		
a. corridors connecting alternative exits	E 20 S_a[2]	FD 20 S[2]
b. dead-end portions of corridors from the remainder of the corridor.	E 20 S_a[2]	FD 20 S[2]
8. **Any door within a cavity barrier**	E 30	FD 30
9. **Any door that forms part of the enclosure to a protected entrance hall or protected landing in a flat**	E 20	FD 20
10. **Any door that forms part of the enclosure:**		
a. to a place of special fire hazard	E 30	FD 30
b. to ancillary accommodation in care homes (see paragraph 2.44).	E 30	FD 30
11. **In a dwellinghouse:**		
a. between a dwellinghouse and a garage	E 30 S_a[2]	FD 30 S[2]
b. forming part of the enclosures to a protected stairway in a single family dwellinghouse	E 20	FD 20
c. within any fire resisting construction in a dwellinghouse not described elsewhere in this table.	E 20	FD 20

NOTES:

1. Classified in accordance with **BS EN 13501-2**. National classifications do not necessarily equate with European classifications, therefore products cannot typically assume a European class unless they have been tested accordingly.
2. Unless pressurisation techniques that comply with **BS EN 12101-6** are used, these doors should also comply with one of the following conditions.
   a. Have a leakage rate not exceeding $3m^3$/m/hour (from head and jambs only) when tested at 25Pa under **BS 476-31.1**.
   b. Meet the additional S_a classification when tested to **BS EN 1634-3**.

D

Appendix D: Methods of measurement

Occupant number

D1 The number of occupants of a room, storey, building or part of a building is either of the following.

a. The maximum number of people it is designed to hold.

b. In buildings other than dwellings, the number of people calculated by dividing the area of a room or storey(s) (m^2) by a floor space factor (m^2 per person) such as given in Table D1 for guidance.

D2 Counters and display units should be included when measuring area. All of the following should be *excluded*.

a. Stair enclosures.

b. Lifts.

c. Sanitary accommodation.

d. Any other fixed part of the building structure.

Table D1 Floor space factors[1]

Type of accommodation[2][3]	Floor space factor (m^2/person)
1. Standing spectator areas, bar areas (within 2m of serving point), similar refreshment areas	0.3
2. Amusement arcade, assembly hall (including a general purpose place of assembly), bingo hall, club, crush hall, dance floor or hall, venue for pop concerts and similar events and bar areas without fixed seating	0.5
3. Concourse or queuing area[4]	0.7
4. Committee room, common room, conference room, dining room, licensed betting office (public area), lounge or bar (other than in (1) above), meeting room, reading room, restaurant, staff room or waiting room[5]	1.0
5. Exhibition hall or studio (film, radio, television, recording)	1.5
6. Skating rink	2.0
7. Shop sales area[6]	2.0
8. Art gallery, dormitory, factory production area, museum or workshop	5.0
9. Office	6.0
10. Shop sales area[7]	7.0
11. Kitchen or library	7.0
12. Bedroom or study-bedroom	8.0
13. Bed-sitting room, billiards or snooker room or hall	10.0
14. Storage and warehousing	30.0
15. Car park	Two persons per parking space

NOTES:

1. As an alternative to using the values in the table, the floor space factor may be determined by reference to actual data taken from similar premises. Where appropriate, the data should reflect the average occupant density at a peak trading time of year.
2. Where accommodation is not directly covered by the descriptions given, a reasonable value based on a similar use may be selected.
3. Where any part of the building is to be used for more than one type of accommodation, the most onerous factor(s) should be applied. Where the building contains different types of accommodation, the occupancy of each different area should be calculated using the relevant space factor.
4. For detailed guidance on appropriate floor space factors for concourses in sports grounds refer to *Concourses* published by the Football Licensing Authority.
5. Alternatively the occupant number may be taken as the number of fixed seats provided, if the occupants will normally be seated.
6. Shops excluding those under item 10, but including: supermarkets and department stores (main sales areas), shops for personal services, such as hairdressing, and shops for the delivery or collection of goods for cleaning, repair or other treatment or for members of the public themselves carrying out such cleaning, repair or other treatment.
7. Shops (excluding those in covered shopping complexes but including department stores) trading predominantly in furniture, floor coverings, cycles, prams, large domestic appliances or other bulky goods, or trading on a wholesale self-selection basis (cash and carry).

Travel distance

D3 Travel distance is measured as the shortest route. Both of the following should be observed.

a. If there is fixed seating or other fixed obstructions, the shortest route is along the centre line of the seatways and gangways.

b. If the route includes a stair, the shortest route is along the pitch line on the centre line of travel.

Width

D4 Width is measured according to the following.

a. For a **door (or doorway)**, the clear width when the door is open (Diagram D1).

b. For an escape route, either of the following.

   i. When the route is defined by walls: the width at 1500mm above finished floor level.

   ii. Elsewhere: the minimum width of passage available between any fixed obstructions.

c. For a **stair**, the clear width between the walls or balustrades. On escape routes and stairs, handrails and strings intruding into the width by a maximum of 100mm on each side may be ignored. Rails used for guiding a stair-lift may be ignored, but it should be possible to park the lift's chair or carriage in a position that does not obstruct the stair or landing.

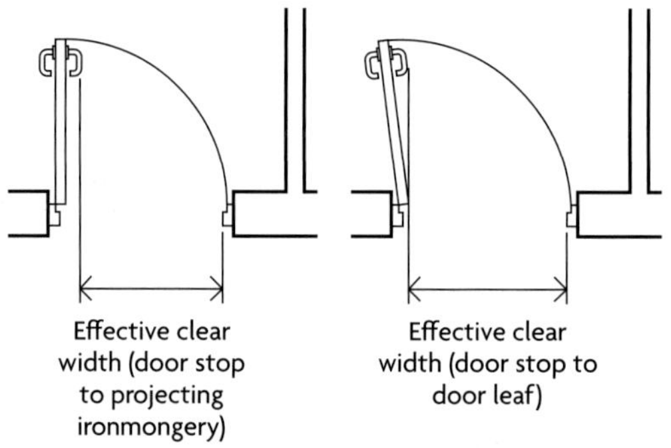

Diagram D1 Measurement of door width

Building dimensions

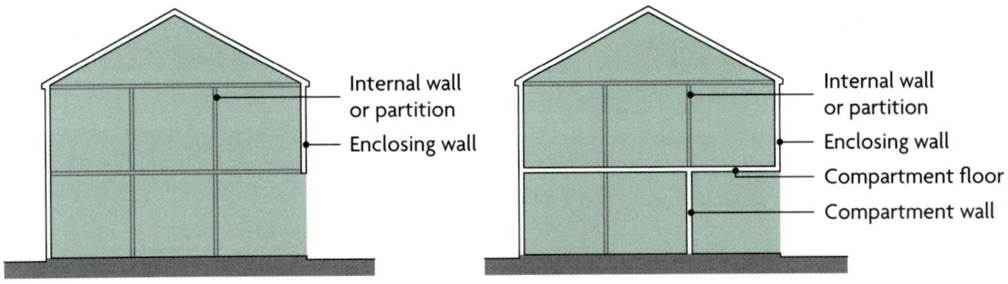

a. **Cubic capacity of a building**

b. **Cubic capacity of compartments or separated part of a building**

In every case measure the volume contained by all of the following.

a. Under surface of roof.

b. Upper surface of lowest floor.

c. Inner surface of enclosing walls. When there is not an outer enclosing wall, measure to the outermost edge of the floor slab.

The measured volume should include internal walls and partitions.

Diagram D2 Cubic capacity

a. Surface area: roofs and rooflights
In each case measure the visible area

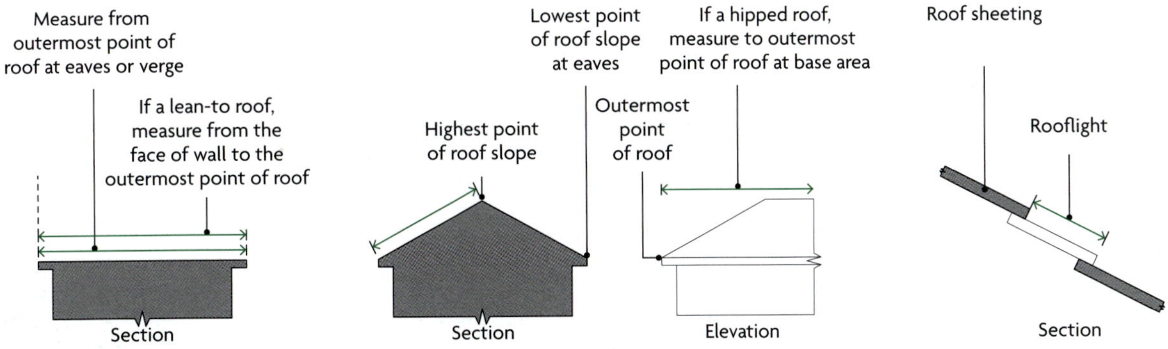

i. Flat or monopitch roof

ii. Double pitch roof

iii. Rooflight

b. Floor area:
Room, garage, conservatory or outbuilding, measure to inner surface of enclosing walls

c. Floor area:
Storey, part or compartment, measure to inner surface of enclosing walls and include internal walls and partitions

When there is not an outer enclosing wall, measure to the outermost edge of the floor slab

Diagram D3 Area

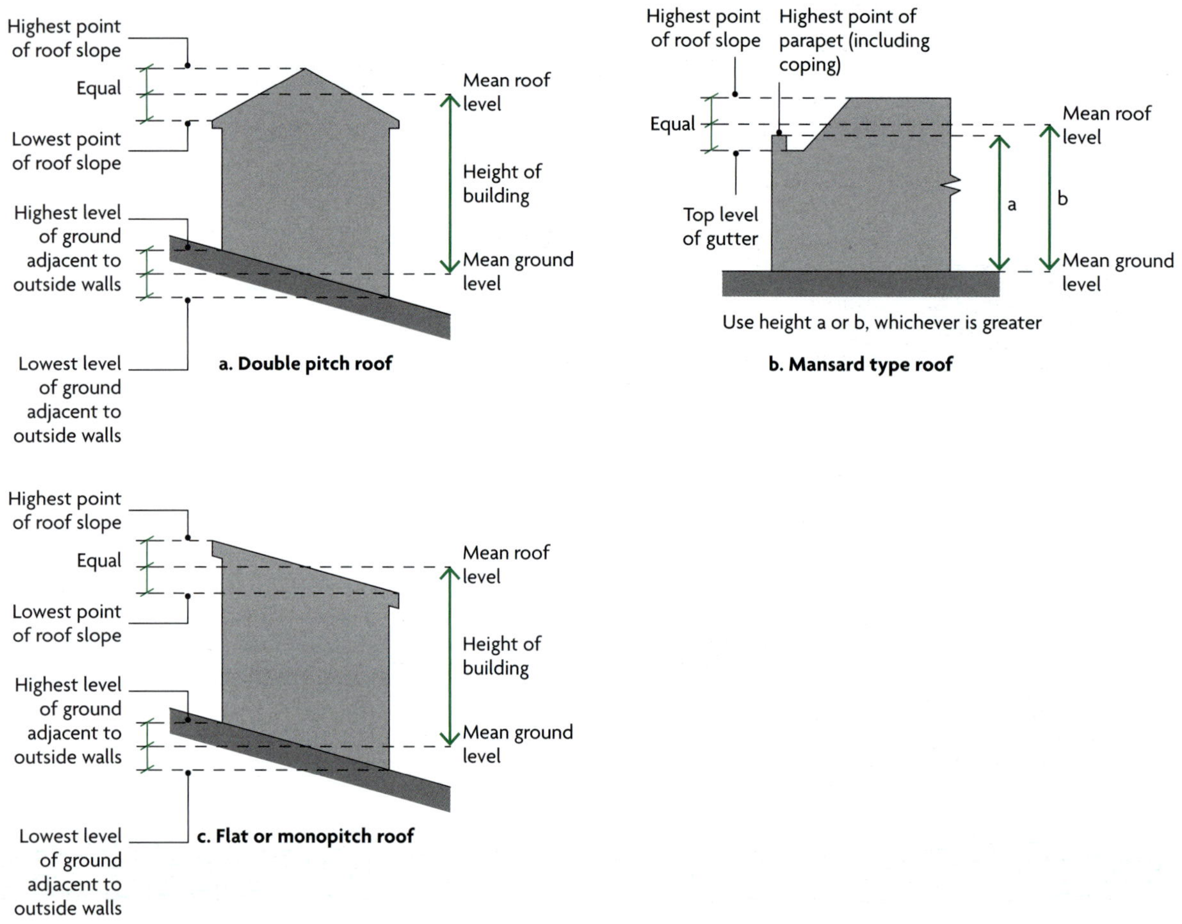

Diagram D4 Height of building

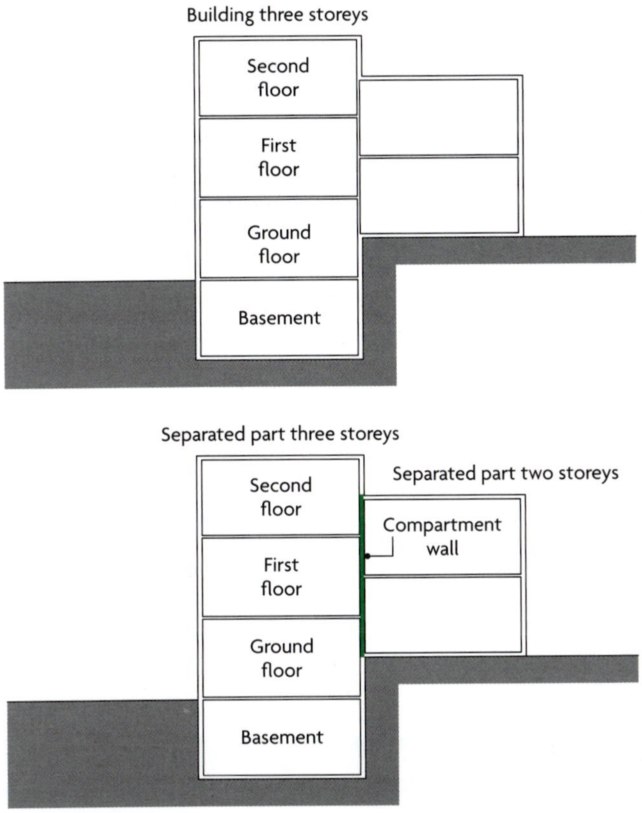

To count the number of storeys in a building, or in a separated part of a building, count only at the position which gives the greatest number and exclude any basement storeys.

NOTES:

1. In assembly buildings (purpose group 5), a gallery is included as a storey, but not if it is a loading gallery, fly gallery, stage grid, lighting bridge, or any gallery provided for similar purposes, or for maintenance and repair.
2. In other purpose group buildings, galleries are not counted as a storey.
3. For the definition of basement, see Appendix A.

Diagram D5 Number of storeys

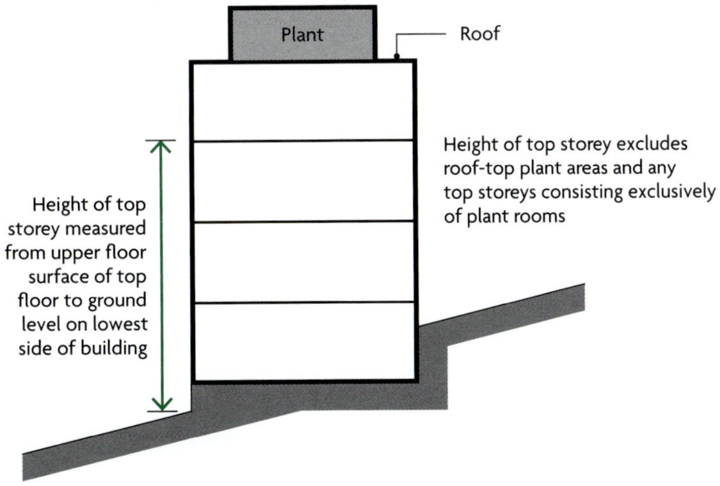

Diagram D6 Height of top storey in building

Free area of smoke ventilators

D5 The free area of a smoke ventilator should be measured by either of the following.

a. The declared aerodynamic free area in accordance with **BS EN 12101-2**.

b. The total unobstructed cross-sectional area (geometric free area), measured in the plane where the area is at a minimum and at right angles to the direction of air flow (Diagram D7).

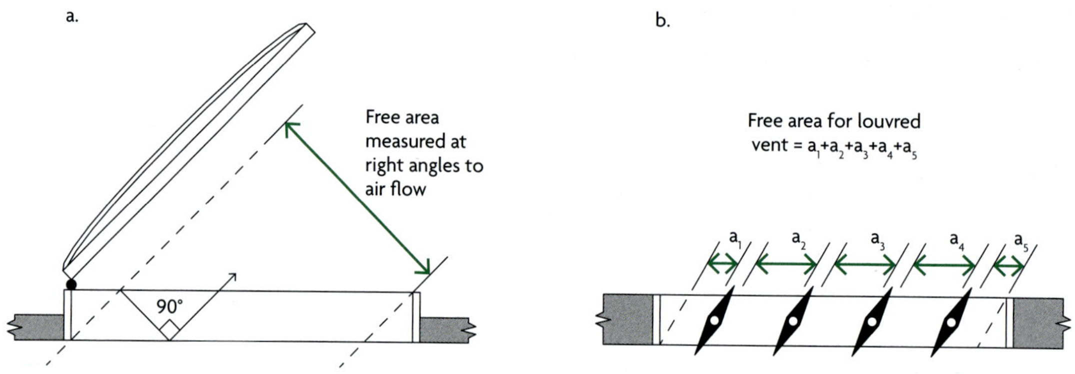

Diagram D7 Free area of smoke ventilators

Appendix E: Sprinklers

Sprinkler systems

E1 Sprinkler systems installed in buildings can reduce the risk to life and significantly reduce the degree of damage caused by fire within a building.

E2 Further recommendations for the provision of sprinklers are provided in the following sections:

Volume 1 – Dwellings

Functional requirement	Paragraph	Title
B1	2.6	Dwellinghouses with two or more storeys more than 4.5m above ground level
B1	2.23	Loft conversions
B1	3.21	Internal planning of multi-storey flats
B3	Table B4	Minimum periods of fire resistance
B3	7.4	Sprinklers
B4	11.15	Unprotected areas and fire resistance – portal frames
B4	11.21	Methods for calculating acceptable unprotected area – sprinkler systems
B5	15.7	Provision of firefighting shafts

Volume 2 – Buildings other than dwellings

Functional requirement	Paragraph	Title
B1	2.46	Residential care homes – sprinkler systems
B1	3.21	Width of escape stairs – phased evacuation
B1	5.46	Shop store rooms
B3	7.7	Raised storage areas
B3	Table 8.1	Maximum dimensions of building or compartment
B3	Table B4	Minimum periods of fire resistance
B3	8.14	Sprinklers
B4	13.16	Unprotected areas and fire resistance – portal frames
B4	13.22	Methods for calculating acceptable unprotected area – sprinkler systems
B5	17.8	Location of firefighting shafts
B5	18.11	Provision of smoke outlets – mechanical smoke extract

Design of sprinkler systems

E3 Where required, sprinkler systems should be provided throughout the building or separated part, unless acting as a compensatory feature to address a specific risk. They should be designed and installed in accordance with the following.

a. For residential buildings, the requirements of **BS 9251**.

b. For non-residential buildings, or residential buildings outside the scope of **BS 9251**, the requirements of **BS EN 12845**, including the relevant hazard classification together with additional measures to improve system reliability and availability as described in Annex F of the standard.

NOTE: Any sprinkler system installed to satisfy the requirements of Part B of the Building Regulations should be provided with additional measures to improve system reliability and availability and is therefore to be regarded as a life safety system. However, there may be some circumstances in which additional measures to improve system reliability and availability specified in Annex F of **BS EN 12845** are inappropriate or unnecessary.

E4 If the provisions in a building vary from those in this document, sprinkler protection can also sometimes be used as a compensatory feature.

BS 9251 makes additional recommendations when sprinklers are proposed as compensatory features.

Water supplies and pumps

E5 For non-residential sprinkler systems designed and installed to **BS EN 12845**, water supplies should consist of either of the following.

a. Two single water supplies complying with clause 9.6.1, independent of each other.

b. Two stored water supplies meeting all of the following conditions.

   i. Gravity or suction tanks should satisfy all the requirements of clause 9.6.2(b), other than capacity.

   ii. Any pump arrangements should comply with clause 10.2.

   iii. In addition to meeting the requirements for inflow, either of the following should apply.

   - The capacity of each tank should be at least half the specified minimum water volume of a single full capacity tank, appropriate to the hazard.
   - One tank should be at least equivalent to half the specified water volume of a single full capacity tank, and the other shall not be less than the minimum volume of a reduced capacity tank (see clause 9.3.4) appropriate to the hazard.

   The total capacity of the water supply in (iii), including any inflow for a reduced capacity tank, should be at least that of a single full holding capacity tank that complies with Table 9, Table 10 or clause 9.3.2.3, as appropriate to the hazard and pipework design.

E6 For the systems described in paragraph E5, both of the following apply if pumps are used to draw water from two tanks.

a. Each pump should be able to draw water from either tank.

b. Any one pump, or either tank, should be able to be isolated.

The sprinkler water supplies should not be used as connections for other services or other fixed firefighting systems.

Appendix F: Standards referred to

European Standards

NOTE: All the British and European Standards can be purchased at the following address: https://shop.bsigroup.com/. Alternatively access to the British and European Standards may be gained at public reference libraries.

BS EN 54 Fire detection and fire alarm systems

- **BS EN 54-7** Smoke detectors. Point smoke detectors that operate using scattered light, transmitted light or ionization [2018]
- **BS EN 54-11** Manual call points [2001]

BS EN 81 Safety rules for the construction and installation of lifts

- **BS EN 81-20** Lifts for the transport of persons and goods. Passenger and goods passenger lifts [2014]
- **BS EN 81-58** Examination and tests. Landing doors fire resistance test [2018]
- **BS EN 81-72** Particular applications for passenger and goods passenger lifts. Firefighters lifts [2015]

BS EN ISO 306 Plastics. Thermoplastic materials. Determination of Vicat softening temperature (VST) [2013]

BS EN 520 Gypsum plasterboards. Definitions, requirements and test methods [2004 + A1 2009]

BS EN 1125 Building hardware. Panic exit devices operated by a horizontal bar, for use on escape routes. Requirements and test methods [2008]

BS EN 1155 Building hardware. Electrically powered hold-open devices for swing doors. Requirements and test methods [1997]

BS EN 1366 Fire resistance tests for service installations

- **BS EN 1366-2** Fire dampers [2015]
- **BS EN 1366-8** Smoke extraction ducts [2004]

BS EN 1634 Fire resistance and smoke control tests for door and shutter assemblies, openable windows and elements of building hardware

- **BS EN 1634-1** Fire resistance test for door and shutter assemblies and openable windows [2014 + A1 2018]
- **BS EN 1634-2** Fire resistance characterisation test for elements of building hardware [2008]
- **BS EN 1634-3** Smoke control test for door and shutter assemblies [2004]

BS EN 1993-1-2 Eurocode 3. Design of steel structures. General rules. Structural fire design [2005]

BS ISO 3864-1 Graphical symbols. Safety colours and safety signs. Design principles for safety signs and safety markings [2011]

BS EN 12101 Smoke and heat control systems

- **BS EN 12101-2** Natural smoke and heat exhaust ventilators [2017]
- **BS EN 12101-3** Specification for powered smoke and heat control ventilators (Fans) [2015]
- **BS EN 12101-6** Specification for pressure differential systems. Kits [2005]

BS EN 12845 Fixed firefighting systems. Automatic sprinkler systems. Design, installation and maintenance [2015]

BS EN 13238 Reaction to fire tests for building products. Conditioning procedures and general rules for selection of substrates [2010]

BS EN 13501 Fire classification of construction products and building elements

- **BS EN 13501-1** Classification using data from reaction to fire tests [2018]
- **BS EN 13501-2** Classification using data from fire resistance tests, excluding ventilation services [2016]

BS EN 13501-3 Classification using data from fire resistance tests on products and elements used in building service installations: fire resisting ducts and fire dampers [2005 + A1 2009]

BS EN 13501-4 Classification using data from fire resistance tests on components of smoke control systems [2016]

BS EN 13501-5 Classification using data from external fire exposure to roof tests [2016]

BS EN 14306 Thermal insulation products for building equipment and industrial installations. Factory made calcium silicate (CS) products. Specification [2015]

BS EN 14604 Smoke alarm devices [2005]

BS EN 15102 Decorative wall coverings. Roll and panel form [2007 + A1 2011]

BS EN 15650 Ventilation for buildings. Fire dampers [2010]

BS EN 15725 Extended application reports on the fire performance of construction products and building elements [2010]

BS EN 50200 Method of test for resistance to fire of unprotected small cables for use in emergency circuits [2015]

British Standards

BS 476 Fire tests on building materials and structures

BS 476-3 Classification and method of test for external fire exposure to roofs [2004]

BS 476-6 Method of test for fire propagation for products [1989 + A1 2009]

BS 476-7 Method of test to determine the classification of the surface spread of flame of products [1997]

BS 476-8 Test methods and criteria for the fire resistance of elements of building construction [1972]

BS 476-11 Method for assessing the heat emission from building materials [1982]

BS 476-20 Method for determination of the fire resistance of elements of construction (general principles) [1987]

BS 476-21 Methods for determination of the fire resistance of loadbearing elements of construction [1987]

BS 476-22 Methods for determination of the fire resistance of non-loadbearing elements of construction [1987]

BS 476-23 Methods for determination of the contribution of components to the fire resistance of a structure [1987]

BS 476-24 Method for determination of the fire resistance of ventilation ducts [1987]

BS 476-31.1 Methods for measuring smoke penetration through doorsets and shutter assemblies. Method of measurement under ambient temperature conditions [1983]

BS 2782-0 Methods of testing. Plastics. Introduction [2011]

BS 3251 Specification. Indicator plates for fire hydrants and emergency water supplies [1976]

BS 4422 Fire. Vocabulary [2005]

BS 4514 Unplasticized PVC soil and ventilating pipes of 82.4mm minimum mean outside diameter, and fittings and accessories of 82.4mm and of other sizes. Specification [2001]

BS 5255 Specification for thermoplastics waste pipe and fittings [1989]

BS 5266-1 Emergency lighting. Code of practice for the emergency lighting of premises [2016]

BS 5395-2 Stairs, ladders and walkways. Code of practice for the design of helical and spiral stairs [1984]

BS 5438 Methods of test for flammability of textile fabrics when subjected to a small igniting flame applied to the face or bottom edge of vertically oriented specimens [1989]

BS 5446-2 Fire detection and fire alarm devices for dwellings. Specification for heat alarms [2003]

BS 5499 Graphical symbols and signs

BS 5499-4 Safety signs. Code of practice for escape route signing [2013]

BS 5499-5 Safety signs, including fire safety signs. Signs with specific safety meanings [2002]

BS 5839 Fire detection and fire alarm systems for buildings

BS 5839-1 Code of practice for system design, installation, commissioning and maintenance of systems in non-domestic premises [2017]

BS 5839-2 Specification for manual call points [1983]

BS 5839-3 Specification for automatic release mechanisms for certain fire protection equipment [1988]

BS 5839-6 Code of practice for the design, installation, commissioning and maintenance of fire detection and fire alarm systems in domestic premises [2019]

BS 5839-8 Code of practice for the design, installation, commissioning and maintenance of voice alarm systems [2013]

BS 5839-9 Code of practice for the design, installation, commissioning and maintenance of emergency voice communication systems [2011]

BS 5867-2 Fabrics for curtains and drapes. Flammability requirements. Specification [2008]

BS 5906 Waste management in buildings. Code of practice [2005]

BS 7157 Method of test for ignitability of fabrics used in the construction of large tented structures [1989]

BS 7273 Code of practice for the operation of fire protection measures

BS 7273-4 Actuation of release mechanisms for doors [2015]

BS 7346-7 Components for smoke and heat control systems. Code of practice on functional recommendations and calculation methods for smoke and heat control systems for covered car parks [2013]

BS 7974 Application of fire safety engineering principles to the design of buildings. Code of practice [2019]

BS 8214 Timber-based fire door assemblies. Code of practice [2016]

BS 8313 Code of practice for accommodation of building services in ducts [1997]

BS 8414 Fire performance of external cladding systems

BS 8414-1 Test method for non-loadbearing external cladding systems applied to the masonry face of a building [2015 + A1 2017]

BS 8414-2 Test method for non-loadbearing external cladding systems fixed to and supported by a structural steel frame [2015 + A1 2017]

BS 8519 Selection and installation of fire-resistant power and control cable systems for life safety and fire-fighting applications. Code of practice [2010]

BS 9251 Fire sprinkler systems for domestic and residential occupancies. Code of practice [2014]

BS 9252 Components for residential sprinkler systems. Specification and test methods for residential sprinklers [2011]

BS 9990 Non automatic fire-fighting systems in buildings. Code of practice [2015]

BS 9991 Fire safety in the design, management and use of residential buildings. Code of practice [2015]

BS 9999 Fire safety in the design, management and use of buildings. Code of practice [2017]

Appendix G: Documents referred to

Legislation

(available via www.legislation.gov.uk)

Education Act 1996

Gas Safety (Installation and Use) Regulations 1998 (SI 1998/2451)

Lifts Regulations 1997 (SI 1997/831)

Pipelines Safety Regulations 1996 (SI 1996/825)

Prison Act 1952

Safety of Sports Grounds Act 1975

Regulatory Reform (Fire Safety) Order 2005 (SI 2005/1541)

Commission Decision 2000/367/EC of 3 May 2000 implementing Council Directive 89/106/EEC

Commission Decision 2000/553/EC of 6 September 2000 implementing Council Directive 89/106/EEC

European Parliament and Council Directive 95/16/EC

Other documents

Publications

Association for Specialist Fire Protection (ASFP)
(www.asfp.org.uk)

ASFP Red Book – *Fire-Stopping: Linear Joint Seals, Penetration Seals and Cavity Barriers*, Fourth Edition

ASFP Grey Book – *Volume 1: Fire Dampers (European Standards)*, Second Edition

ASFP Blue Book British Standard version – *Fire Resisting Ductwork, Tested to BS 476 Part 24*, Third Edition

ASFP Blue Book European version – *Fire Resisting Ductwork, Classified to BS EN 13501 Parts 3 and 4*, First Edition

Ensuring Best Practice for Passive Fire Protection in Buildings, Second Edition [2014]

Building Research Establishment Limited (BRE)
(www.bre.co.uk)

BRE report (BR 135) *Fire Performance of External Thermal Insulation for Walls of Multi-storey Buildings*, Third Edition [2013]

BRE report (BR 187) *External Fire Spread: Building Separation and Boundary Distances*, Second Edition [2014]

BRE Digest 208 *Increasing the Fire Resistance of Existing Timber Floors* [1988]

BRE report (BR 274) *Fire Safety of PTFE-based Materials Used in Buildings* [1994]

Department for Communities and Local Government
(www.gov.uk/government/publications/fire-performance-of-green-roofs-and-walls)

Fire Performance of Green Roofs and Walls [2013]

Department for Education
(www.dfes.gov.uk)

Building Bulletin (BB) 100: *Design for Fire Safety in Schools* [2007]

Department of Health
(www.dh.gov.uk)

Health Technical Memorandum (HTM) 05-02: *Firecode. Guidance in Support of Functional Provisions (Fire Safety in the Design of Healthcare Premises)* [2015]

HTM 88: *Guide to Fire Precautions in NHS Housing in the Community for Mentally Handicapped (or Mentally Ill) People*

Door and Hardware Federation (DHF) and Guild of Architectural Ironmongers (GAI)
(www.firecode.org.uk)

Hardware for Fire and Escape Doors [2012]

Door and Shutter Manufacturers' Association (DSMA)
(www.dhfonline.org.uk)

Code of Practice for Fire Resisting Metal Doorsets [2010]

Fire Protection Association (FPA)
(www.thefpa.co.uk)

RISCAuthority Design Guide for the Fire Protection of Buildings [2005]

Football Licensing Authority
(www.flaweb.org.uk/home.php)

Concourses [2006]

Glass and Glazing Federation (GGF)
(www.ggf.org.uk)

A Guide to Best Practice in the Specification and Use of Fire-resistant Glazed Systems [2011]

Health and Safety Executive (HSE)
(www.hse.gov.uk)

Safety Signs and Signals: The Health and Safety Regulations 1996. Guidance on Regulations, L64 [2015]

HM Prison and Probation Service (HMPPS)
(www.hmppsintranet.org.uk/uploads/HMPPSFireSafetyDesignGuide.pdf)

Custodial Premises Fire Safety Design Guide

Passive Fire Protection Federation (PFPF)
(http://pfpf.org/pdf/publications/guide_to_uailoft.pdf)

Guide to Undertaking Assessments in Lieu of Fire Tests [2000]

Sports Grounds Safety Authority
(https://sgsa.org.uk/)

Guide to Safety at Sports Grounds [2007]

Steel Construction Institute (SCI)
(https://steel-sci.com)

SCI Publication P288 *Fire Safe Design: A New Approach to Multi-storey Steel-framed Buildings,* Second Edition [2006]

SCI Publication P313 *Single Storey Steel Framed Buildings in Fire Boundary Conditions* [2002]

Timber Research and Development Associations (TRADA)
(www.trada.co.uk)

Timber Fire Resisting Doorsets: Maintaining Performance under the New European Test Standard [2002]

Index

A

B

C

D

E

F

G

H

I

J

K

L

M

Q

R

S

T

U

V

W